LIVESTOCK PRODUCTION AND RURAL DEVELOPMENT IN INDIA

LIVESTOCK PRODUCTION AND RURAL DEVELOPMENT IN INDIA

Edited by:
Birendra Kumar Mishra

BIOTECH
2012
BIOTECH BOOKS

ISBN 978-81-7622-224-0

Published by : **BIOTECH BOOKS**
4762-63/23, Ansari Road, Darya Ganj,
New Delhi - 110 002
Phone: +91-011-23262132
E-mail: biotechbooks@yahoo.co.in

Laser Typesetting : **Sushil Kumar**

Printed at : **Chawla Offset Printers**
Delhi - 110 052

PRINTED IN INDIA

Dedicated
to
my beloved teacher
Prof. R.K. Yadav

FOREWORD

Animal husbandry and agriculture are the twin occupations which are being practiced in the rural households since ancient times. Livestock sector plays an important role in Indian economy and is an important sub-sector of Indian Agriculture. The contribution of livestock to GDP of India was 4.70 percent in 2004-05 at 1999-2000 prices which is in increasing trend in comparison to the contribution of agriculture. The overall growth rate in livestock sector is steady and is around 4-5% and this has been achieved despite the fact that investment in this sector was not substantial. The bulk of our livestock are in the hands of landless labourers and marginal farmers. So the progress in this sector results in balanced development of the rural economy particularly in reducing the poverty amongst the weaker section. The rural women play a significant role in Animal Husbandry and are directly involved in most of the operations relating to feeding, breeding, management and healthcare of the livestock. Livestock provides stability to family income and these are the best insurance against the vagaries of nature due to drought, famine and other natural calamities. Major part of the livestock population is concentrated in the marginal and small size of holdings. Livestock plays an important and vital role in providing nutritive food to families both in rural and urban areas. Bullock power continues to be the main source of draught power for agricultural operations and transport of agricultural products to nearby markets and is likely to remain so for a long time to come. Further, agricultural productions get valuable organic manure provided by the livestock. So the impact of livestock keeping in the lives of rural poor in India is enormous, and it is particularly true in the north-eastern hill region of our country. Livestock production is an integral and inseparable part of north-eastern states in particular. It plays a vital role in the socio-economic condition of the farmers. Percentage of rural population in NEH region ranges from 85% in Meghalaya to 96% in Arunachal Pradesh, whereas the all India average is 80%. Most of these rural people are doing agriculture by occupation. A major portion of the farmers are basically small and marginal farmers. The total land availability for agriculture in this region is 2270 million hectors and out of which 70% is located in Assam, 8% in Manipur and 7.5% in Tripura. The area available for other states is very negligible. Therefore, farmers find animal husbandry to be a better alternative. Moreover, the tribal people of NEH states are exclusively non-vegetarian in their dietary habits and they prefer meat and meat products. Tribal people constitute 79% of the population in Arunachal Pradesh, 88.6% in Nagaland and 80% in Meghalaya. There is no taboo of taking any type of meat and their most preferred meat is pork. Therefore, importance of livestock production including piggery has given special emphasis in this region. Amongst various livestock species, traditionally people are rearing cattle (*Bos indicus*), fowl (*Gallus domesticus*), pig (*Sus domesticus*), duck (*Anas platyrhynchos*), goat (*Capra hircus*) and dog (*Canis familiaris*), and these are the common and conventional enterprises in rural as well as urban areas throughout the north-eastern states. Domesticated animals are mainly of *deshi* / local variety or breed with some exceptional cases where crossbred animals are being reared. Besides, lack of superior germplasm for higher productivity, the other constraints of livestock production in India and in the NEH region in particular are non-availability of livestock feed (both in quality and quantity), more disease occurrence due to heavy rainfall and high humidity, lack of organized slaughter house for scientific slaughter and hygienic packaging of meat and meat products for proper marketing through out the region, and lack of scientific knowledge in relation to package of practices of various livestock species.

The present book entitled **'Livestock Production and Rural Development in India'** is designed to ameliorate the above mentioned problems of livestock production by incorporating various livestock production strategies in India with special reference to north-eastern hill region viewed by eminent scientists in this field. I am confident that the effort of Dr Birendra Kumar Mishra in bringing out this volume will be appreciated by the students, teachers, policy planners and all others concerned in this field. I wish him all success in his enterprise.

Dr. L. Mandal
Professor & Ex-Dean
Faculty of Agriculture
Department of Animal Science
BCKV,
Nadia, West Bengal-741252, India

(L. Mandal)

Dated-27.06.2011

ACKNOWLEDGEMENTS

I am at loss of words to bring out my deep sense of veneration and gratitude to Dr.L.K.Jha, Professor, School of Human and Environmental Sciences,.North-Eastern Hill University, Shillong for his valuable guidance and suggestion at the initial stage of my carrier and also in pursuit the present work.

The academic encouragement also received from Prof. Omprakash, Ex. Director, NIRD, Guwahati, Prof. S.S.Khare, Ex. Pro-vice-chancellor, NEHU, Tura Campus, Tura, Meghalaya and Prof. R.K. Yadava my most beloved teacher and also my supervisor at Ph.D. level.

I am highly indebted to the family members of North-Eastern Hill University, Tura Campus and to my colleagues of the Department and my friends of the campus for encouraging me in my academic job at all times.

Special thanks go to my parents for their moral support and continuous encouragement throughout my life.

I wish to record a word of appreciation to Biotech Book Publishers, New Delhi for accepting the manuscript and bringing out this book in a graceful manner.

Last but not least, it is my pleasure to record and acknowledge the cheerful cooperation of my wife Dr. (Mrs.) Shalini Mishra and both of my sons Shreyansh and Deeptansh.

02nd, August, 2011 **Birendra Kumar Mishra**

INTRODUCTION

North-Eastern region of India consists of the contiguous seven sisters (Assam, Meghalaya, Manipur, Nagaland, Tripura, Mizoram and Arunanchal Pradesh) states and Sikkim. It is ethnically distinct from the other states of the country. More than 2000 Km. of boundary is shared with other countries. These states comprised of 72 district; 39,946 villages and 245 towns. The region is underdeveloped in terms of livestock production than rest of the country. Very few organized industries are available in this region. This region known for its unique culture, handicrafts, martial arts and scenic value. Problems including insurgency, unemployment drug addiction and lack of infrastructure. Since beginning of the economic liberalization in 1990, studies have shown that this region is lagging behind the others in terms of development.

The main agricultural activity of farmers in North – East region of the country is the mixed crop-livestock system characterized with low inputs and outputs, making poor farmers suffer from decreasing sustainability. Livestock, especially pig and poultry, is a key livelihood among the farming rural families and provides food for consumption, sources of income and capital and socio-cultural needs. There is an increasing demand for livestock products in the region. The region procures 10 millon eggs and 2500 tones of milk every month. These suggest that livestock development is a large opportunity to improve the livelihoods among poor farmers in the region.

The growth of livestock sector is very slow in this region than the national level. However, a significant proportion of landless laborers, small and marginal farmers have access to livestock resources and the acceleration in the growth of livestock sector in the region offers significant opportunities for household income augmentation and employment generation. It also performs an important input functions in terms of contributing draught power and dung to crop production. The basic factors responsible for growth in the livestock rearing in North-East region are unavailability of best breed, feeds and fodder, poor road facilitates and lack awareness etc. These factors should be addressed to accelerate the development of livestock sector in this part of the country, which is an important source of livelihood for millions of poor people. Moreover, among various agricultural enterprises, livestock production has more income redistributive on household level and is very effective in reducing rural income inequality. Therefore, supportive technical, institutional and policy initiatives for improvement of breeds, feed availability, disease control, food safety and private investment are further required to expand production and improve productivity. Attention should also be paid to improve the viability of small farms by improving their accessibility to both input and output markets.

The present work is an attempt in the aforesaid direction. The book is collection of research article and papers on various aspects of Livestock Production and Rural Development in India with special reference to North-East region. We have divided the whole book into two sections. A-section-wise highlight of this book is presented below:

The first section deals with- "**Major Issues Relating to Livestock and Rural Development with special reference to North-East region**" In this section the paper entitled Livestock: A Basic component of Integrated Farming System in North-East Region presented by **Mr. M.Islam,** he studied that how livestock may be the basic component of integrating farming system. He concluded that the efficient integration of enterprises taking livestock as a basic component of farming system rather than traditional knowledge based integration; it may enhance the sustainability of the farm. **S.C.Srivastava** in a paper entitled Agrarian Reforms: Historical Perspective and Neglected Issues observed that our recent agricultural policy reforms in the country should be seen as a long term process of policy changes initiated since independence and can not be separated from the rural development policies of the country. This has helped because of the some neglected issues primarily concerned with small and marginal farmers of the country.**Chandrashekhar Choubey and B.K.Mishra** in his article reviewed the status of Livestock Development in Meghalaya, they critically described the present status of livestock in the state and what are the possible challenges and issues for the development of livestock. **S.Malik, D.K.Pandey and M.Datta** attempt to analyze the sustainable rural poultry production in North-East region. They observed that the majority of the birds in the region are non-descript, in which growth and production is very slow. They also suggested in his paper that how profitable poultry faming can be achieved through scientific management of poultry farms.

D.P.Singh and D.C.Rai articulated Infertility and Sterility Problems in Dairy animals. In their study, they said since the economics of dairy is not based on milk production but largely depend on the full function of reproductive efficiency. To explore the full production potential there should be high degree of reproductive capability in the animals. Organic Farming Scope and Limitations presented by **Sunil Kumar,** suggestes that the low external inputs based Indian Dairy sector having better opportunity to convert organic dairy production, since majority of the farmers in our country are organic farmers not by choice but by default. **Manoj Kr Tiwari,** examined in his paper titled –Rationality of Ancient Cattle Management Practices in the Era of Present Context, he elaborated that the ancient wisdom and knowledge is sensitive and comprehensive and can be adapted to also changing climatic condition. The identified key areas of livestock research focus should be given urgently, because the growth and development of livestock is directly related to community in terms of nutritional and food security as well as for employment. A study on application of GIS in dairy industry have been presented by **Deepak Bhagat and Shweta Priyamvada** ,they suggested that the GIS may lead to a better and improved supply chain management in dairy industry with some suitable examples. **Rajendra Kumar Pandey** presented a review article entitled –Prospects of Dairy Development and Indigenous Dairy

Products in North-Eastern region he has given a brief historical background along with future prospects and challenges for the development dairy and indigenous milk products in the region. In another paper presented by **J.U.Ahmad** on financing issues in agricultural sector: an implication, he suggested on the basis of his study that there should be proper recovery of loan amount from loaner in order to revamp the performance of credit delivery institutions. This would lead the nourishment of a new and active agricultural sector in the study area. **K. Hriiyio Kayina** suggested in his study entitled –Supply Chain Management in the Poultry Meat Industry that this lowers the transaction cost and increased the margin money, because of the many activities and aspects involved .It demands a multidisciplinary approach and trade relation.

Section two depicts the livestock development and livelihood- we included seventeen papers in this section, first paper of this section contributed by **Ashish. K. Makwana** he highlighted the present status of World Trade Organization and Indian Dairy Industry in a very brief manner. **Biswajit Paul** examined in his study entitled Participation of Women in Livestock Production in India, he focused in the paper that the involvement and efficiency of women in the concerned field throughout the world, India in general and North-East India in particular. **Professor L.Mandal** presented a paper entitled- Scientific Poultry Farming, in this paper he recommended the ways and means of profitable poultry farming in a very nut cell. **Smt.Ajanta Deb (Kar)** presented a paper entitled-Livestock Production and Women-A Compatible Pair, inthis paper she addressed the role and involvement of women in livestock management activities in different production system all over the world. She also recommended that women need to be empowered through awareness and training etc.after all they can take equal participation with men in decision making in this system. In another paper **Biswajit Paul** again highlighted the positive and negative impact of livestock in the environment in a very effective way. Poultry Farming as a means of Livelihood for small Farmers in Garo Hills presented by **B.K.Mishra** he focused that it is a very important component of livelihood especially in rural community. This also may be a backyard venture for most of the farmers and can fulfill the house hold demand of the farmers in terms of nutrition. He highlighted the strategies about scientific management system for better performance of poultry farms. **Ch.Nandkishore Singh and Victoria Devi** reviewed an article entitled Present Status of Mithun Rearing in Manipur; they observed that this is a very unique animal suitable for livestock production system in North-Eastern hilly region of the country. In this regard they highlighted the package and practices of Mithun rearing in very effective way. **Anindita D.Sangma** in her paper on Livestock by-products and its Biological Industrial Importance: In this paper she focused that what are the possible by-products can be generated from slaughter houses and how these can be utilized in food, feed and pharmaceutical industries. Economic Evaluation of Settled and Shifting Cultivation along with their Farming System in Hill Zones of Assam presented by **D.C. Kalita and B.C Bhowmic.**They found in his study that the output and input ratio was higher in settled cultivation than the shifting cultivation. Meat consumption Pattern among major tribes in Nagaland have been studied by **N.K. Patra, M.N. Odyuo, Ch.Victoria Devi and S. Mondal** .They have taken five popular sources of meat in the study area and found the preference of meat according to taste, as per scale value it was more or less same and the result reveals that the dog meat losing its popularity. **S.K.Pyne and G.Samanta** conducted an experiment on neuro muscular deformities in lead induced chick embryos, they found the affected chocks dies within 21 days after hatching. They also noticed that both the Lead compounds were determental to chick embryos; out of which lead acetate were more lethal than nitrate. In another paper entitled –information communication technology for entrepreneurship in rural dairy farming. He attempted to identify how ICT contribute the growth and sustainability of this sector and also focused on potential ICT application to this sector. **S. Naskar and Nilotpal Ghosh** presented a paper entitled- an approach for improving livelihood security through sustainable pig production. In this paper the authors highlighted the importance of livestock production in NEH region; particularly the piggery has given special emphasis. They also suggested some important key points for better management of piggery units in this region. A-review on status of duck development in Manipur submitted by **K.Rashbery Singh**,He reviewed the status of duck farming, present status and future prospects in a very nice manner . The last paper submitted by **Ashish .K.Makwana** on Indian Women in Dairy Business-A Case study of Chitoor District in Andhra Pradesh. In this paper his finding was that the percentage of women has risen more than 20%, but their participation in cooperative management committees is negligible as paid positions.

In this edited volume the theory and practice of livestock production and rural development has been set out in very simple language in a very clear and precise form. It represents the fruit of one year of labour and gives a broad picture of many problems facing the livestock and rural development workers today.

I hope that this useful volume of authentic information on the above subject will meet the urgent and practical requirement of students, progressive farmers, educationist and extension workers.

B K Mishra
Assistant Professor
Department of RDAP
North-Eastern Hill University,
Tura Campus
Tura-794002, Meghalaya

LIST OF CONTRIBUTORS

A.K. Mishra: Jawahar Lal Nehru Krishi Vishwa Vidyalaya College of Agriculture, Tikamgarh (M.P.) 472001

A.K.Gaur: Jawahar Lal Nehru Krishi Vishwa Vidyalaya College of Agriculture, Tikamgarh (M.P.) 472001

Smt. Ajanta Deb (Kar): Department of Economics, Don Bosco College, Tura, email – ajantakardeb@yahoo.com

Anindita D.Sangma: Department of Rural Development and Agricultural Production North-Eastern Hill University, Tura Campus,Tura-794002, Meghalaya

Ashish. K. Makwana : Dairy Business Management Department, SMC College of Dairy Science, Anand Agricultural University Anand-388110 Gujarat, e-mail: ashish_gau@rediffmail.com Mobile No: 09898273887

B.K.Mishra: Department of RDAP, North-Eastern Hill University, Tura Campus, Tura-794002, Meghalaya

Biswajit Paul: Don Bosco College,Tura,Meghalaya, email – bptura@yahoo.co.in

C.S.Choubey: Department of Animal Husbandry &Dairying, Kulbhasker Ashram P.G.College, Allahabad (U.P)

Ch. Victoria Devi: Department of Agricultural Extension, Bidhan Chandra Krishi Viswavidyalaya, PO - Krishi Viswavidyalaya, Mohanpur, Nadia, West Bengal – 741252, E-mail: ch_venus2005@yahoo.com

D. K. Pandey: Department of fisheries Extension, College of Fisheries Science, Lembucherra – 799 210, Tripura (West)

D. P. Singh: Department of Animal Husbandry and Dairying Institute of Agricultural Sciences, Banaras Hindu University, Varanasi (U.P.)

D. C. Rai: Department of Animal Husbandry and Dairying Institute of Agricultural Sciences, Banaras Hindu University, Varanasi (U.P.)

D.C. Kalita and B.C. Bhowmick : Department of Rural Development and Agricultural Production North-Eastern Hill University, Tura Campus, Tura-794002, Meghalaya

Deepak Bhagat & Shweta Priyamvada*: Department of Management, NEHU, Tura Campus (dip19bhagat@gmail.com) * Student, Department of Geography, Cotton College, Guwahati University

G Samanta: Animal Science Department of ASEPAN, Palli Siksha Bhavana Visva-Bharati, Sriniketan, W.B.

Jaynal Ud-din Ahmed : Department of Management, North Eastern Hill University Tura Campus, Chandmari, Tura - 794002, West Garo Hills, Meghalaya, India E-mail: jaynaluahmed@yahoo.co.in / juahmednehu@gmail.com

J.B. Singh: Jawahar Lal Nehru Krishi Vishwa Vidyalaya College of Agriculture, Tikamgarh (M.P.) 472001

K.Rashbery Singh, Deputy Director (Instruction) Central Agriculture University, Iruisemba, Imphal-795004, Manipur, e-mail: cau_dde@rediffmail.com

K.Hriiyio Kayina: Research Scholar Department of Management, North-Eastern Hill University, Tura Campus Tura-794002, Meghalaya

Lalmohan Mandal: Department of Animal Science, BCKV, Mohanpur, Nadia, West Bengal.

Manoj Kumar Tiwari: M.K.Tiwari,Agriculture Extension Officer, Zila Parishad, Collectorate Campus, Bhilwara (Rajasthan)-311001, e-mail: mkt_sept@yahoo.com

Md. Mokidul Islam: Krishi Vigyan Kendra, Tura, ICAR Research Complex for NEH Region Sangsanggre, Dobasipara-794 005 West Garo Hills (Meghalaya) e-mail: mislam01d@yahoo.co.in

M. N. Odyuo: Department of Agricultural Extension, School of Agricultural Sciences and Rural Development, Nagaland University, Nagaland – 797106, E-mail: nk_patra08@yahoo.in

Nilotpal Ghosh: Department of Animal Science, Bidhan Chandra Krishi Viswavidyalaya, PO - Krishi Viswavidyalaya, Mohanpur, Nadia, West Bengal – 741252; E-mail:gnilotpal@yahoo.com

Nirmal Kumar Patra: Department of Agricultural Extension, School of Agricultural Sciences and Rural Development, Nagaland University, Nagaland – 797106, E-mail: nk_patra08@yahoo.in

Rajendra Kumar Pandey: Department of Animal Husbandry & Dairying, In-Charge Dairy Farm Institute of Agricultural Sciences, Banaras Hindu University, Varanasi-Uttar Pradesh-221 005

Ram Priti Mani Tripathi: Lecturer, B.R.D.P.G. College Deoria (U.P.)

S. C Srivastava: Department of RDAP North Eastern Hill University, Tura Campus Tura-794002, Meghalaya

S. K. Pyne: Animal Science Department of ASEPAN, Palli Siksha Bhavana Visva-Bharati, Sriniketan, W.B.

S. Malik: ICAR Research Complex for NEH Region Tripura Centre, Lembucherra – 799 210, Tripura (West)

M. Datta: ICAR Research Complex for NEH Region Tripura Centre, Lembucherra – 799 210, Tripura (West)

Sagar Mondal: Department of Agricultural Extension, Bidhan Chandra Krishi Viswavidyalaya, PO - Krishi Viswavidyalaya, Mohanpur, Nadia, West Bengal – 741252, E-mail: sagarmondal.bckv@gmail.com, ch_venus2005@yahoo.com

Sunil Kumar: College of Forestry and Hill Agriculture, G.B.P.U.A. &T., Hill Campus, Ranichauri, Tehri Garhwal, Uttarakhand, 249199, E-mail: sunilgbpuat@gmail.com

S.Naskar: Indian Veterinary Research institute, Eastern Regional Station, 37, Belgachia Road, Kolkata -700037; E-mail: s_naskar@rediffmail.com

Sunil Doley: Senior Scientist Division of Poultry Science ICAR Research Complex for NEH Region, UMROI Road, UMIAM, Meghalaya-793 103

* * *

CONTENTS

Section- I

Major Issues Relating to Livestock and Rural Development with Special Reference to North-East Region

Section –II

Livestock Development and Livelihood

1

LIVESTOCK: A BASIC COMPONENT OF INTEGRATED FARMING SYSTEM IN NORTH EASTERN REGION

Md. Mokidul Islam
Subject Matter Specialist (Agronomy),
Krishi Vigyan Kendra, Tura, ICAR Research Complex for NEH Region,
Sangsanggre, Dobasipara-794 005, West Garo Hills (Meghalaya)
E-mail: mislam01d@yahoo.co.in

ABSTRACT

Our Indian agriculture depends on agriculture livestock. In Indian agriculture livestock is play vital role in farm economy as an integral part of the farming system. They provides farm power, rural transport, manures, cooking fuels, milk and meat having major role in rural economy by providing income and infrastructure to small land holding farmers and other weaker section of the society. Today livestock sector play major role in the global economy. Each develops countries tries to provide enough milk and meat to each people.Livestock sector has been playing an important sub-sector of Indian economy and important sub-sector of Indian agriculture. The contributions of livestock to GDP decrease from 5.22% in 1999-2000 to 4.36% in 2004-2005 at current prices. According to estimate GDP from livestock sector at current prices was about Rs. 935 billion during 1999-2000. These values rose to Rs. 1239 billion during 2004-2005 with 24.72% share in agriculture and allied GDP. This sector also played an important role in supplementing nutritive food rich in animal protein to general public and supplementing family income and generating gainful employment in the rural sector particularly among the landless and small marginal farmers and women. With its large livestock population has vast potential for meeting the growing needs of millions of families in respect of livestock product such as meat and milk. It also provides animal proteins and various types of raw materials for industrial used.

Introduction

The North Eastern Region (NER) comprises the states of Arunachal Pradesh, Assam, Meghalaya, Manipur, Mizoram, Nagaland, Sikkim and Tripura. The total geographical area of 26.3 million hectares and a population of 40 million (2001 Census) representing 8 and 4 per cent of the total area and population of the country, respectively. The North Eastern Hill (NEH) region which excludes the state of Assam consists of about 70% of the geographical area is hilly and mountainous with the elevation ranging from 0 to 200 m above MSL in Tripura to more than 5000 m above MSL in Arunachal Pradesh. The climate varies from tropical to alpine. The NEH Region is characterised by different terrain, wide variations in slopes and altitude, land tenure systems and cultivation practises. The agricultural production system in the region is mostly rainfed, monocropped and at subsistence level in the plains, valleys, foothills and terraced slopes. Use of local varieties, low consumption of chemicals, low moisture retention capacity of upland soil, undulating topography, low temperature, no drainage in valley land during monsoon and lack of irrigation facilities during winter months, low sunshine hours along with traditional management practices have resulted into low productivity of crops and low cropping intensity in the region. Problems of high rainfall (12.1% of the country's total precipitation), soil acidity, aluminium toxicity in upland and iron toxicity in valley land have added to the problem of low agricultural productivity in the region. The soils of the region are usually rich in organic matter and acidic to strongly acidic (pH 4.5-5.0) in reaction and broadly classified into five orders of Inceptisols, Entisols, Alfisols, Ultisols and Molisols. Most of the hill soils are shallow in depth except in valleys and plateaus. Soil erosion is a major problem in the region due to steep slopes and faulty land use practices. Slash and burn agriculture (Jhum cultivation) is still predominantly practiced in almost all the NE states of hill region except Sikkim, on steep slopes with reduced low cycle of 2-3 years as against 10-15 years in the past. This system has lead to accelerated erosion, silting of reservoirs and low productivity of land and ecological imbalance well. The region is bestowed with rich natural resources of soil, water and agro-climate, ttraditional dependence of rural tribal population in livestock for livelihood and non-vegetarian nature of food habits of the people in NEH states, thin population density and village leadership pattern make it easy to introduce, popularise and adoption of integrated farming system with livestock as a basic component of the system.

Livestock: A basic component of IFS

Livestock including poultry is an integral part of agriculture and contributes significantly to rural economy in the country. While agriculture contributes about 30% of the total output of GDP, livestock including poultry constitutes about 25% of the agricultural output. Rearing of livestock and poultry is an age- old practice in India with no exception to the North Eastern States particularly in Meghalaya. About 80% of its contribution to Gross National Product is from Cattle and Buffaloes with Sheep. Goats, Pigs and Poultry covering the balance 20%. Livestock production has a much larger contribution to manpower employment than agriculture while eggs, milk and meat production help in stepping up crop production through the availability of cash to the livestock farmer round the year which is utilised for purchase of seed, manure and fertilisers for agriculture operation. Establishment of dairy, poultry, sheep &goat or pig farms helps to solve unemployment problems and provide valuable protein, vitamin and mineral rich materials like eggs, milk and meat. Organic manure from livestock and poultry enrich the soil for higher production of crop, vegetable and flower as well as horticultural produce.

However, the food security of our country is likely to be threatened in the near future due to high population pressure. According to a recent estimate the food grain requirement of India would be 380 million tonnes and 450 million tonnes in the years of 2025 and 2050 respectively, with the present rate of population growth. Use of high yielding varieties, extension of irrigation facilities and greater application of fertilisers and pesticides contributed to a radical enhancement of food grain production in the mid 60's of the last century which is known as green revolution. The green revolution concentrated only on single enterprises; crop more specially wheat and rice. Although later on developments in other enterprises like dairy, fishery etc took place, but in an isolated manner. The only way out is to diversify the economic activities in an interdependent and integrated manner at micro level within the available resources. Similarly, in North Eastern states especially in Meghalaya, agricultural operations having limitations due to its topography, climatic situation and socio-economic conditions claiming only about 10% of the total land for cultivation, livestock and poultry provide the only alternative avocation the villagers fall upon for a subsidiary living. Application of modern technologies has brought a high hope for commercial livestock and poultry farming for full time occupation for a decent living. Hence, the integrated diversified farming system with livestock as a basic component would provide employment round the year to improve the purchasing power of the farming community in the north eastern hill regions of the country.

Status of Livestock in NE Region

The North Eastern Region posses a large number of different livestock species (Table 1) wherein agriculture and animal husbandry are complementary and not competitive to each other and the livestock completely dependent on crop based feeding system.

Table 1: Livestock population in North Eastern States (million)

Species	NE States	All India
Cattle	11.488	185.181
Buffalo	0.840	283.446
Sheep	0.227	61.469
Goat	4.366	124.358
Pig	3.82	13.52
Yak	0.016	0.065
Mithun	0.254	0.278

Source: 17th Livestock Census of India, 2003

Table 2: Density of livestock population per Sq.Kms in Meghalaya

Sl. No.	District	Cross-bred Cattle	Indig-enous Cattle	Buffa-loes	Pig	Sheep	Goat	Poultry	
								Fowls	Ducks
1	East Khasi Hills.	1.93	19.22	0.3	31	4.17	11.96	69.74	0.51
2	Ri-Bhoi.	2.75	16.52	1.83	7.72	0.03	5.46	76.21	1.22
3	West Khasi Hills.	0.1	17.75	1.23	5.82	1.69	8.37	44.38	-
4	Jaintia Hills.	0.13	26.09	0.52	9.97	0.02	5.68	54.22	0.49
5	East Garo Hills.	0.4	40.72	1.34	13.96	-	6.65	124.74	4.52
6	West Garo Hills.	1.03	54.84	4.07	18.15	0.07	13.37	131.47	8.19
7	South Garo Hills	0.04	22.64	0.75	9.45	0.01	10.1	78	1.16
	MEGHALAYA	**0.79**	**28.41**	**1.5**	**13.11**	**1.03**	**8.72**	**79.16**	**2.26**

Scope and prospects

The NE state particularly in hilly tribal states of the region has a great scope and prospect for livestock development which would help in the integration for the integrated farming system. These are

- Traditional dependence of rural tribal population in livestock for livelihood;
- A food habit of the people is mostly non-vegetarian in nature.

- Absence of any taboo for consumption of chicken meat, pork or beef.
- Absence of sentiments for slaughter of livestock including cattle for meat supply.
- Readiness of the farmers to adopt modern technologies for breeding, feeding, management and disease control and their willingness to avail of Govt. services.
- Quality of cows in the NE State particularly in Meghalaya is quite good and given the due support these animals could adopt themselves with the developmental process to produce quicker results as compared to other States.
- Availability of favourable markets within the NE State in general and Meghalaya in particular.

On the contrary, there are some "**minus points**" as well and these are:

- Dairy farming was not that popular amongst the local inhabitants in NEH Region earlier but mostly restricted to Nepalese immigrants who took it up as a commercial venture.
- Consumption of liquid milk in rural areas of NEH Region is not popular with consequent lower demand of milk as a whole.
- In the NE Region, particularly in NEH region a very high proportion of forestland is owned by public and as such, grassland and fodder development activities will be difficult to make a break through without people's participation.
- Some of the essential infrastructures for A.I. Services, Diagnostic services, availability of bio-logical like vaccines etc. and training of farmers in different areas etc. are inadequate. As such, considerable time will be required to set them up prior to taking up massive programmes of livestock development in NE H Region in the country.

- Above all locational disadvantages with consequent hazards in communication system create problems for input supplies and marketing in other North Eastern Hill region excluding plains of Assam in general and Meghalaya in particular.

Strategy

The strategy for livestock development in NE Region would be called for

- Qualitative improvement in all classes of domestic livestock e.g., Cattle Buffalo, Sheep, Pigs, Poultry, Ducks and Rabbit to increase the overall productivity in these animal.
- In doing so, the aim will be to increase the per capita availability of milk, egg and meat.
- This is to be achieved through creation of adequate infrastructure for animal production, health coverage, processing of animal products and organisation of an efficient marketing system to match the production so envisaged.
- Production of fodder and balanced livestock feed will receive high priority to make animal production activities cost effective and remunerative to the farmers.
- Greater participation of farmers will be sought for and for this purpose, organisation of farmers co-operatives, involvement of N.G.Os in various activities will be encouraged.
- Training is the basic requirement for entrepreneurs taking up livestock farming on commercial basis and as such Entrepreneurship Development Programme (E.D.P) will be one of the prime thrust areas in the plan.

Feeds and fodder

Straws, stovers and other agricultural by products would continue to be major input as livestock feed for ruminants. Unfortunately, the existing technologies for improving digestibility of straws have remained mostly confined to laboratories and not used by the farmers mainly because of high costs and cumbersome processes involved. Biotechnological techniques which can develop recombinant microbes to digest straws, neutralize lignin and its by products, and release carbohydrates through a solid state fermentation process, shall be researched so as to make available energy for livestock feeding.

Productivity of livestock is dependent on availability of quality feed and fodder in requisite quantities. Production of coarse grains has gone down over the years. There is large shortage of green fodder and concentrates. Efforts would be made to make available required quantities of Cereals and oil meals for livestock and poultry sector. Non-conventional animal feed resources shall be exploited to make available protein for livestock feeding. Throughout the North Eastern Region, cost and availability of feed are probably the most important constraint to increase livestock production since feed costs are the major part of production costs that accounts for 60-80 percents. The gap between demands and availability of feed (Table 3) resources are big and in alarming situation in the country as a whole and North Eastern States in particular.

Table 3: Demands and availability of livestock feed and fodder (million tonnes)

Types of feed	NEH Region			All India		
	Demand	Availability	Gap	Demand	Availability	Gap
Dry roughages	12.00	12.00	-	799	551	248
Green fodder	74.60	34.40	40.20	558	430	128
Concentrate	5.13	0.90	4.23	117	62	55

Source: JJ Gupta (2007)

Fodder production in NE Region

Adoption of scientific fodder production and conservation by utilising the local germplasm and improved seeds and cuttings obtained from reputed organisation, in the homestead, fallow or unutilised land and adoption of Agro-Forestry would bring in better availability situation of feeds and fodder for the livestock for optimum production. Promising and suitable fodder crops should be grown based on sustained yield of fodder, fast coverage, adaptability in soil condition and palatability amongst different species of livestock in the region. Forages from the major portion of feed of cattle during June to September, which is replaced with shrubs, herbs and other tree leaves during lean season. Some improved perennial grasses have an excellent production potential in the region. Promising variety of improved perennial grasses such as setaria (*Setaria sphacelata)*, congosignal (*Brachiaria rosenensis)*, guinea(*Panicum maximum)*, para (*Brachiaria mutica)*, and broom (*Thysanolaena maxima)* for pasture development to economize milk, beef and chevon production. Besides these, *stylosanthese* and groundnut fodder variety are promising perennial legumes. Their yield performances however, reduced during winter season due to moisturestress and cold. Rice bean, cowpea, soybean performed well and suitable annual legumes for low, medium and high altitude upto 1500 m above msl. Navaro, Parari, Exbucklandia and Mulberry have been found suitable fodder tree for feeding to rabbit, goat and dairy animals (Sahoo *et al.,* 2000). The majority of tree leaf fodder are high in crude protein and desirably low to medium in crude fibre content. It has an important role in feeding during lean season when availability of grasses is scanty.

Concept of Integrated Farming System(IFS)

The Integrated farming system is based on the concept that " there is no waste" and "waste is only a misplaced resource which can become a valuable material for another product" (Sarangi *et al,* 2007). It is particularly relevant to benefit the rural poor. All the components of agriculture like crop-livestock-fish-forestry-horticulture are integrated in a complementary way. Thus, the integrated farming system (IFS) may provide the environmentally sustainable and economically viable technology that encompasses rational utilization of available resources of the region. The emphasis in such system is on optimizing resource utilization rather than maximization of individual element in the system. The suitable IFS model can be shown in fig.1

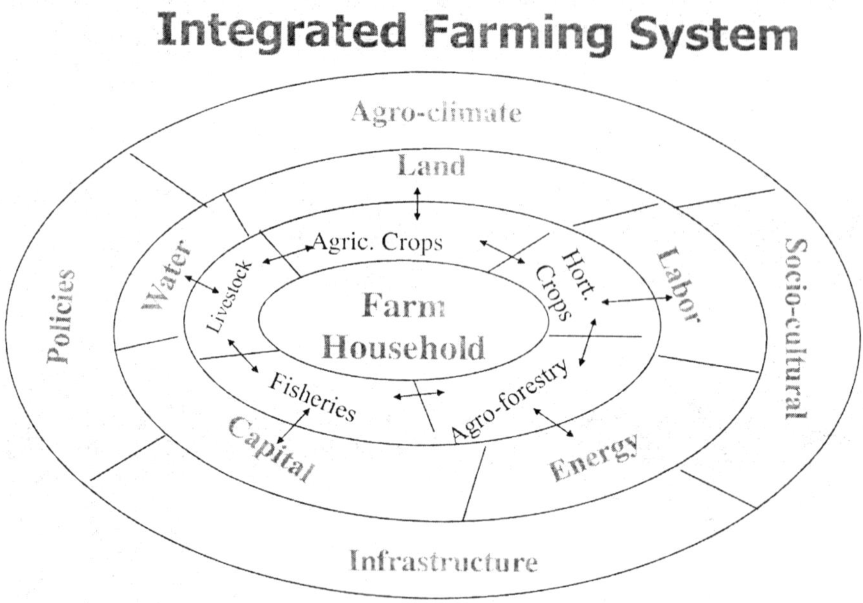

Fig. 1: Schematic view of Integrated Farming System (IFS)

Definition of IFS

The Integrated Farming System (IFS) is the integration of two or more enterprises like crop, livestock, poultry, fishery, horticulture, agro-forestry, bee-keeping, agro-industry and allied activities etc for each farm according to the availability of resources to sustain and satisfy as many necessities of the owner as it is possible. It is possible even for a resource poor farmer and for him integrated system is more important to maximise the profit through proper utilization of scarce resources in a farming system approach.

The IFS can remove all the constraints by not only solving most of the existing economic and even ecological problems, but also provide the needed means of production such as fuel, fertilizer and feed, besides increasing productivity many-fold. It can turn all those existing disastrous farming systems into economically viable and ecologically balanced systems that will not only alleviate poverty, but also can even eradicate this source completely.

Integration of enterprises

In agriculture, crop husbandry is the main activity. The income obtained from cropping is hardly sufficient to sustain the farm family throughout the year. Assured regular cash flow is possible when cropping is combined with other enterprises especially livestock. Judicious combination of enterprises, keeping in view of the environmental conditions of a locality will pay greater dividends. At the same time, it will also promote effective recycling of residues/wastes.

Choice of enterprise

The basic points that are to be considered while choosing appropriate enterprise in Integrated Farming System (IFS) are:

- Soil and climatic features of an area/locality
- Resource availability with the farmer
- Present level of utilization of resources
- Return/income from the existing farming system
- Economics of proposed integrated farming system
- Farmer's management skill
- Social customs prevailing in the locality

Component of IFS

Livestock is the best complementary enterprise with cropping, especially during the adverse years. Installation of a biogas plant in crop-livestock system will make use of the wastes, at the same time provides valuable manure and gas for cooking and lighting. In a wetland farm, there are greater avenues for fishery, duck farming and buffalo rearing. Utilizing the rice straw, mushroom production can be started. Pigs are the unique components that can be reared with the wastes which are unfit for human consumption. In rainfed farming, sheep and goat rearing form an integral part of the landscape. Agroforestry(Agri-Horti-Silvi, Silvi-horti, Agri-horti-pastoral etc) are other activities which can be included under dryland hill agro-ecosystem conditions. The components of IFS in NEH Region are:

- Hedgerow intercropping
- Livestock rearing, including poultry and duckery, and composite fish culture
- Horticultural crops
- Agroforestry including forage crops
- Fish farming
- Medicinal herbs and lesser known wild edibles
- Variety of crops including seasonal vegetables
- Liquid manure production
- Mushroom production
- Vermiculture
- Apiculture

IFS for NEH Region

The North Eastern Hill Region of India are unique in their physiographic characteristics of land, varied agro-climatic conditions, land tenure system and cultivation practices from the rest of the country. The region falls under high rainfall zone and is characterised by hilly terrain with wide variation in elevation from 100 m to 5500 m. Majority of the farmers in the region are maintain dairy cattle and major source of green fodder comes from lopping of fodder trees and locally available grasses. Crops and cropping patterns of such land uses will differ depending on the type of enterprise such as milk, beef, mutton, wool, pork and poultry production. Combination of cultivated varieties of perennial grasses, legumes, shrubs and trees can extend availability of green fodder upto February at low altitude, thereby shortening the requirement of conserved fodder for lean season feeding. Land upto 100% slope having soil depth greater than 1 m can be used for agri-horti-silvipastoral system. This system has the potentiality of maintaining 1.18 livestock unit/ha of land use in NEH Region. The integrated farming systems have been tested at ICAR Research Complex for NEH Region are:

Cow-fish: A cattle population of 6 nos. would be ideal for fertilizing a pond of 1 ha area. A stocking density of 7500 nos. of fish fingerlings per ha would be optimum

Pig-fish: A population of 40 nos. of pigs would be ideal for fertilizing a pond of 1 ha area. A stocking density of 7500 nos. of fish fingerlings per ha would be optimum. In this system, fish production of about 2200 kg per ha per year can be realised

Duck-fish: A population of 300 ducks would be ideal for fertilizing a pond of 1 ha area. A stocking density of 7500 nos. of fish fingerlings per ha would be optimum. In this system, fish production of 2000 kg per ha per year can be realised.

Poultry-fish: A population of 400 - 500 nos. of poultry birds would be ideal for fertilizing a pond of 1 ha area. A stocking density of 7500 nos. of fish fingerlings per ha would be optimum. In this system, fish production of 2000 kg per ha per year can be realised

Fish cum horticulture: The pond dykes can be used for growing wide varieties of vegetables and fruit crops. Vegetables like cauliflower, cabbage, frenchbean, chillies, cucumber, pumpkin and tomato can be grown. Fruit crops like guava, litchi, banana, peach, oranges, mango and papaya perform well.

Intensive Integrated Farming Systems (IIFS): IIFS refers to sustainable production on the one hand and livelihood security on the other, wherein all the components of agriculture, horticulture, forestry, livestock, fishery can be integrated in a complimentary way besides soil conservation measures, vermicompost, mushroom production apiculture and liquid manure preparation, etc. The suitable tree-crop combinations can find the place in IIFS.

Table 4: Monetary input/output pattern (Lac/yr) of IIFS

Intensive Integrated Farming Systems(IIFS)	Total input	Total output	Output/ Input ratio with labour	Output/ Input ratio without labour
Poultry-crop-fish-duck-horticulture-hedgerow	1.06	1.67	1.58	2.24
Crop-fish-poultry-MPTs	0.60	0.91	1.51	2.12
Crop-fish-goat-bamboo-MPTs	0.594	0.92	1.55	2.40
Crop-fish-pig-hedgerow-MPTs-vermiculture	0.772	1.10	1.42	1.86
Crop-fish-dairy-mushroom-liquid manure-broom-horticulture-vermiculture	1.70	2.99	1.76	2.48
Crop and fishery (Control)	0.32	0.35	1.09	1.50

Source: Annual Report, ICAR, Umiam

Advantages of IFS

- Increased productivity per unit of land
- Better utilization of resources
- Recycling of farm wastes
- Sustainability of the enterprises in the system
- Employment generation
- Reduction of risk

Conclusions

The necessity is the mother of invention and well established with the integration rather than isolation. Hence, the efficient integration of enterprises taking livestock as a basic component of integrated farming system rather than traditional knowledge based integration which leads to sustainable productivity of a farm.

References:

Annual Report. 2005-08. ICAR Research Complex for NEH Region, Umiam, Meghalaya.

Biswas, C. and Singh, R. 2003.Integrated farming system: An intensive approach. *Intensive agriculture.* pp.22-23.

Gupta,J.J. 2007. Status and Technology of Fodder Production,pp.59-80. (In) *Sustainable Crop Production- for food sufficiency in NE India*, Panwar, A.S., Ngachan, S.V., Munda, G.C. and Das,A. Eds.478p. ICAR Research Complex for NEH Region, Umiam, Meghalaya

Munda,G.C., Das,A., Panwar,A.S. and Ngachan, S.V. 2007. Status and Technology of Crop production, pp 32-58. (In) *Sustainable Crop Production- for food sufficiency in NE India*, Panwar, A.S., Ngachan, S.V., Munda, G.C. and Das,A. Eds.478p. ICAR Research Complex for NEH Region, Umiam, Meghalaya

Sahoo, S.K., Yadav, B.P.S., Gupta, J.J. and Gupta, H.K. 2000. Performance of goat fed exbucklandia (*Symingtonia populnea*) leaves. *Indian Journal of Animal Nutrition, 17*(4): *336-338.*

Sarangi,S.K., Lama,T.D. and Satapathy,K.K. 2007. Evaluation of organic Farming components through Farming System in NEH Region, pp.325-335. (In)*Advances in Organic Farming Technology in India,* Munda, G.C., Ghosh,P.K., Das,A., Ngachan,S.V. and Bujarbaruah,K.M. Eds.585p.ICAR Research Complex for NEH Region, Umiam, Meghalaya

Sarangi,S.K. and Satapathy,K.K. 2007 Crop Production Through Farming System Approach,pp.116-137. (In) *Sustainable Crop Production- for food sufficiency in NE India*, Panwar, A.S., Ngachan, S.V., Munda, G.C. and Das,A. Eds.478p. ICAR Research Complex for NEH Region, Umiam, Meghalaya

* * *

2

AGRARIAN REFORMS: HISTORICAL PERSPECTIVES AND NEGLECTED ISSUES

DR. S C SRIVASTAVA
Professor, Department of RDAP, North Eastern Hill University, Tura campus, Tura- 794002, Meghalaya

ABSTRACT

Recent agricultural policy reforms in India should be seen as a long term process of policy changes initiated since independence and can not be separated from the rural development policies of the country. The policy changes became more focused and market oriented with implementation of new agricultural technology initiated at the behest of Ford Foundation report published in 1958. The first phase of the policies which continued till India became a signatory of WTO. The policies being pursued since the WTO era are not only directly opposed to the earlier polices but have also led to a dichotomy in general economic policy and rural development policies. This has led to an overall deceleration in agriculture growth with increased unsustainability in the post reforms period compared to pre reforms period. Food security has emerged as a primary issue in present context. This has happened mainly because of the neglect of some of the issues primarily issues concerning the small and marginal farmers. Targeting the small and marginal farmers and evolving appropriate institutional arrangement for delivery of inputs and providing marketing outlets, restoring the pace of land reforms and linking it community based land use planning for food security should be the primary aim of future policy reforms in the agriculture sector.

Introduction:

Ever since the process of liberalization started and India became a signatory to World Trade Organization, agriculture has been the burning topic as far as the impact of liberalization and globalization is concerned. While the Government and other official sources continue to highlight the achievements in terms of higher GDP growth rate critics have mainly targeted the agriculture sector for its tardy performance during the post reforms period. From policy point of view the agriculture sector represent one of the largest private sectors in the world where more than 80 million farmers operate with diverse technologies, product varieties and environmental conditions with varying degree of natural and infrastructural support system across the country. Since agriculture is natural resource based, especially soil and water based activity, it also represent the status of country's vast environment. The present paper attempts

a) A Review of reforms in agriculture in the broader context of rural development and emerging policy dichotomy
b) Performance of agriculture sector in the post reforms period vis a vis pre reforms period ,
c) Neglected issues and Emerging Policy Options.

(A)

Rural Development and Agrarian Policy Reforms: A Brief History

Rural Development and Agrarian Reforms were high on economic agenda in Indian policy thinking even much before the independence and persisted ever since the struggle for independence started. While before independence 'social reforms', 'land reforms' and 'agriculture sector reforms' were the key areas of discussion, 'Institutional reforms and land reforms became the main focus of socio- economic policy in the post independent India, especially during the first decade of planning. The rural areas were to develop on the principles of self help and community participation. Consequently "community development programme", "cooperative institutions" and "gram Panchayat" were the corresponding agents of development and change in rural area. The three institutions were supposed to take care of rural people by creating, enhancing and sustaining the rural livelihoods. Agriculture per se during this period was to be developed by following an area based growth strategy.

The first jolt to the philosophy of creating, enhancing and sustaining the rural livelihoods through community endeavours came within even less than ten years of the operation of the model of self-help and people's initiated development. By 1958, Government of India had already started feeling pressure on the foreign exchange front and pangs of hunger becoming more and more pronounced. In 1959, Government of India accepted the Ford Foundation Report (***India's Food Crisis and Steps to Meet It***). So the philosophy of rural livelihoods was to be changed. When the entire country was covered under the community development blocks, the independent Ministry of Community Development was abolished and merged with Ministry of Agriculture in 1960.

The Ford Foundation Strategy had limited perspective of meeting the food crisis and it shifted the focus from community approach of multi-sectoral Rural Development as envisaged by Community Development Programme to entrepreneurial approach of single sectoral agricultural development emphasizing the "production first" philosophy. This approach was based on the hypothesis of Growth first with strong belief in percolation effect. During the decade of 1960s, the country went through unprecedented crisis, especially on international front. Two wars (1962 with China and 1965 with Pakistan), followed by three years of continuous drought disrupted the entire economy as well as country's prestige (mainly because of its dependency on outside world for the food and other essential items, and PL 480 had become the catch word in sixties). Obviously, the agriculture had to be the first priority. While the agriculture sector did pick up in 1970s and mostly in 1980s, the other assumptions failed. The 'growth first' materialized partially in 1970s and mostly in 1980s. In brief, the shift from multi-purpose, multi- sectoral development to single purpose (increased production) single sectoral (agriculture) development meant a change in emphasis from:

a) Community to individual entrepreneurs with the philosophy of "betting on the strong". This also meant 'selective' prioritization of entrepreneurial class in rural society.
b) Organizational and institutional changes to technological upgradation and scientific practices.
c) Integrated co-operative endeavor to private investment in fertilizers, HYV seeds, irrigation, pesticides etc.
d) Introduction of the element of subsidy for encouraging the input use by individual farmers (an issue which attracted maximum attention under the present regime of WTO); and
e) Introduction of pricing structure for foodgrains and other agricultural commodities along with state trading in the form of public procurement and distribution system. This was facilitated by creation of Agricultural Prices Commission (know known as National Commission for Agricultural Costs and Prices), and Food Corporation of India. The former is now under threat of abolition and the latter hot spot for criticisms.

The perusal of the changed approach led to food surplus situation in the country. The decade of 1970s started new chapter in India and for the first time the fruits of modern agricultural practices were felt in the form of self sufficiency and surplus food production. But the associated percolation thesis failed. In fact the reverse happened. The socio- economic consequences of the new agriculture technology and green revolution were quite disturbing from the point of view of the poor in rural areas. Some of the immediate socio economic consequences of the New Agricultural Strategy pointed out by All India Rural Credit Review Committee were-

a) Resumption of land for self-cultivation (for profit motive) resulting into decline of share-cropping and increase in landless and agricultural labour;
b) Increased inter-personal inequality in rural areas leading to agrarian unrest in some parts of the country;
c) Increased inter-regional disparities favoring the agro-climatically privileged areas and States in the country.

Since 1970s therefore, good number of policy measures were initiated to benefit the ecologically constrained regions like desert areas, drought prone areas, hill areas, etc. as well as socially and economically unprivileged classes like small and marginal farmers, landless agricultural labourers, etc. This was a period of integrated approach to agriculture and rural development by putting the thrust on poverty alleviation. At the same time the philosophy of growth by betting on the strong also continued.

During 19980s, agriculture sector in India saw its golden peak. While the policy measures initiated since the adoption of the new agricultural policy continued, the trends in yield growth rate further motivated the Government of India to adopt a region specific and commodity specific approach to agriculture. Depending upon the agro climatic conditions, the country was divided into 15 agro climatic zones during the Seventh Five Year Plan for the purpose of dissemination of technology and organization of the extension works. During the same decade some of the commodity specific missions were started like Special Foodgrain Production Programme (SFPP), National Pulses Development Programme (NPDP), Integrated Oilseeds Policy, Integrated Cotton Development Programme and Special Jute Production Development Programme, etc.. These policies and programmes, during the decade of 1980s, resulted favorable trends and the country witnessed yield based growth rates rather than area based growth rates as observed during 1970s.

Liberalization, Globalization and WTO:

However, most of the policy initiatives listed above became the direct targets of economic reforms, especially after WTO came into existence.

The most important of these issues are a) reduction in farm, seed, fertilizer, irrigation and other input subsidies, b) reduction in food subsidies, and c) rationalization of the pricing structure or Minimum Support Price (MSP) system. Since the beginning of the 1991, the economic policy in India passed through rapid changes, mostly clubbed together in the name of "New Economic Policy" or popularly known as Economic Reforms. Three standard arguments advanced in favour of the Economic Reforms were:

a) It promotes growth by bringing about structural changes in the economy; this is the rationale behind the standard structural adjustment policy (SAP) as advocated by the International Monetary Fund and often thrust upon the developing countries who pursue solutions to their balance of payments (BOP) crisis. The underlying presumption is that the BOP crisis is a reflection of internal economic malfunctioning mainly because of structural and supply side bottlenecks not allowing the economy to operate on the principles of efficiency in resource allocation and competitiveness in the product market.

b) It is pro- poor as growth will trickle down and will take care of poverty. This is essentially based on the premises the relationship between growth and poverty reduction is direct and one way (that is while poverty reduction is possible only with growth, simple poverty reduction policies can not bring about growth).

c) It is pro agriculture sector as it promotes efficiency in the resource use and hence cost and price advantages which ultimately benefit the consumer/ poor. This is directly based on the principles of perfectly competitive market system with the micro economic postulates that under a perfect competitive economy, the price of a product is just equal to its marginal cost which in turn is equal to the average cost. This equality between marginal cost and average cost is a unique feature of free and perfect competitive economy.

The formation of the World Trade Organization was a direct corollary of the last principal. The principles and policies of the WTO are based on Ricardo's concept of comparative advantage. Agriculture as a subject of multilateral negotiation was first introduced during the Uruguay Round of General Agreement on Trade and Tariffs (GATT) in 1986. During the Uruguay Round, for the first time, a framework for Agreement on Agriculture (AoA) was mooted and subsequently became the major agenda item of WTO. As the history of AoA itself suggests, the major concerns of WTO under AoA are-

- Market access
- Domestic support
- Export subsidies
- Trade-Related Intellectual Property Rights

General Economic Reforms: Emerging Policy Dichotomy

While the above developments and reforms were going on in the rural sector, the policy measures with regard to macro economic dimensions of the Indian economy were almost stable till 1979-80 when the major oil crisis broke. For the first time the entire world bowed down to the pressure exerted by the oil oligopolies. India had to take huge assistance from International Monetary Fund to the tune of $ 5.6 billion. The deal was struck by accepting the "Structural Adjustment Programme" of IMF. Thus the current phase of macro economic reforms started in 1979-80. Theoretically, the philosophy of structural adjustment looks at remaining all the supply side bottlenecks from the economy for boosting growth (and employment). Liberalization in the Indian economy thus, actually started much earlier than what is normally perceived. Indian economy witnessed one of most robust performance in every sphere during the 1980s. Not only the trend growth was above 5.5 per cent, but also the growth was well dispersed sectorally. Despite the laudable performance, the 1990s started on very adverse note, not only for India; but for the world as such. While in 1979-80, it was the oil shock alone which shook the world, in 1990, it was coupled with another factor - uni-polarisation of the world. Following

International factors may be linked to set of policy measures taken since 1991 and popularly known as economic reforms as immediate causes-

1. Disintegration of Soviet Union which also meant that India lost its one of biggest trade partner. This factor was missing during the years of structural adjustment (1979-80).
2. The Gulf war of 1991 affecting (a) oil prices and (b) remittance from Middle East. Both the factors acted heavily on BOP crisis.
3. Lowering of India's credit ratings in international market for short term and long term borrowing due the acute political instability.

These factors together resulted in a situation where country had to resort huge borrowings form IMF again, as was done in 1979 and also had to mortgage the gold stocks with the Bank of England. As usual, IMG had its own conditionally of structural adjustment, this time coupled with World Bank's prescriptions for macro economic stabilization.

The kind of policy initiatives taken since 1991 may be summarized as-

a) Trade sector reforms for enhancing globalization,

b) Fiscal sector reforms aiming at reduction in budgetary deficits- both domestic and international. This among other things include reduction and phase wise elimination of subsides of different kind which as discussed earlier were initiated during 1960s to promote the growth of agriculture and infrastructure sector,

c) Industrial sector reforms leading to almost dismantling the industrial licensing system in the country. This also includes selling of (or dis-investment as is popularly known) public sector units into private hands, inviting foreign participation on large scale including in the sectors which were developed and nursed as commanding heights,

d) Agriculture sector reforms virtually leaving agriculture to market forces regardless of the vagaries of nature.

e) Financial sector reforms opening up and inviting foreign financial companies to compete with Indian banking and even insurance sector.

The list could be even larger as any failure of government today is immediately followed by discussion on reforms in that sector. Three common advantages of reforms have been cited by political and economic circles, and by those who advocate it most vehemently- the multilateral development institutions. The three standard arguments are - it promotes growth by bringing about structural changes in the economy; it is pro- poor as growth will take care of poverty; and it is pro agriculture sector as it promotes efficiency in the resource use and hence cost and price advantages which ultimately benefit the consumer/ poor.

This different treatment of rural development which primarily means agricultural development and poverty alleviation and general economic scenario has resulted into a policy dichotomy. The policy dichotomy is very clear from the table 1. While for the rural livelihoods people's participation has the major role to play, for macro economy perfect competition and government has to play the major role. The net impact of this emerging dichotomy of policies have yet to be studied and require a deeper probe in the issue.

Table 1: Guiding Principles and target sectors of reforms

Reforms	Objectives	Guiding Principles	Target sectors	Actors
Rural Sector Reforms	❖Development of agriculture and other sectors ❖Removal of unemployment and poverty ❖Natural resource management	❖People's participation, cooperation ❖Decentralized Governance and decision making ❖Equity and sustainability	❖griculture and other livelihoods	❖People and People's Institutions ❖Bureaucracy ❖Multi lateral and bi- lateral donors
Macro Economic Reforms	❖Macro economic stabilization ❖Economic Growth ❖Balance of Payment ❖Foreign investment ❖Budgetary balance	❖Perfect competition and efficiency ❖Centralized and large scale production ❖Globalization	❖Manufacturing sector ❖Infrastructure ❖Trade ❖Agriculture sector ❖Fiscal Policy	❖Government ❖World bank, IMF ❖Organized manufacturing sector ❖WTO ❖Multilateral corporations ❖National Governments

(B)

Performance of Agriculture since Globalization

The performance of Indian agriculture has to be seen against this background of dichotomy. No wonder that the performance of agriculture since the globalization era has been marred by general tardiness, not only in relative sense but also in absolute sense. The trends are disturbing specially after the WTO regime started. It is therefore imperative that an objective assessment of the performance of agriculture situation in India during the post reforms period be taken and some corrective measures be adopted to face the challenges of globalization and internal issue of food security. In view of this an assessment of the growth performance of Indian agriculture is made below in terms of following parameters-

i) Pattern of Agricultural growth since 1980-81 vis-avis other sectors,
ii) Sustainability of Agricultural growth, and
iii) Trends in foodgrain production and food availability.

Table 2 shows the growth rates of GDP in pre reform period (1981-82 to 1990-91) and post reform period (1991-92 to 2001-2001). Table also shows the sectoral growth rates of GDP during the two periods. Following observations can be made from the table-

i) While the overall growth rate improved in the post reforms period, agriculture sector has seen a setback where the growth rate actually went down. Same is observed about the secondary sector where despite much hue and cry about liberalization, the overall growth rate declined from above seven percent growth rates achieved during the pre-reforms period to below six percent growth rate during the post reforms period. Within the secondary sector the setback is noticed mainly in the manufacturing sector and mining sectors.
ii) Overall, it may be concluded that though on an average the growth rates during the post reforms period have been higher, it did not

depict any significant deviation from the pre reforms period except for the agriculture sector where things have definitely gone worse. The acceleration of growth rate of Indian economy started much earlier during the middle of nineteen eighties, than the initiation of reforms period..

Table 2: Average Annual Sectoral and GDP (at Factor Cost) Growth Rates in Pre and Post Reforms period

Sectors/ Period	1981-82 to 1990-91	1991-92 to 2001-02
Agriculture and Allied Activities	3.22	2.86
Agriculture	3.41	2.87
Industry	**7.14**	**5.62**
Mining and Quarrying	7.68	3.76
Manufacturing	6.92	5.86
Electricity, Gas and water supply	8.30	6.14
Services	**5.86**	**7.28**
Construction	4.27	5.00
Trade , hotels, transport and communications	5.28	7.85
Financing Insurance, Real State and Business services	8.97	8.14
Community, social and personnel services	5.54	6.68
Gross Domestic Product at Factor Cost	**5.39**	**6.18**

Source: National Accounts Statistics, CSO

Sustainability of Agricultural growth:

When analyzed in simple terms of growth, the agriculture sector has shown much unsustainable pattern since liberalization compared to 1980s. Hence most of the changes occurring in the productions sector of agriculture may not necessarily be ascribed to new economic policy as such. Yet the fact remains that growth of agriculture sector during the post reforms period has been fluctuating compared to the decade preceding the reforms period (Table 3). Moreover there has been a general drop in the agricultural growth rate in the face of upsurge in GDP growth rate. More importantly, this little growth was more unsustainable during the post reforms period. During the period in 1981-82 there were three negative growth years amounting to a total of just two percent of negative growth, while in the period 1991-92 to 2001-2002 there were four negative growth years accounting for more than nine percent of negative growth.

Table 3: Growth trends in Agriculture

Annual growth rates in agriculture and allied sectors (at 1993-94 prices)		
Year	**Agriculture**	**Agriculture, Forestry & Fishing**
1981-82	5.6	5.3
1982-83	-0.5	-0.7
1983-84	10.3	9.6
1984-85	1.4	1.5
1985-86	0.7	0.7
1986-87	-0.6	-0.6
1987-88	-1.4	-1.3
1988-89	16.8	15.5
1989-90	0.7	1.5
1990-91	4.4	4.1
Annual Average (1981-82 to 1990-91	3.74	3.56
Total Negative growth	-2.0	-2.6
1991-92	-1.9	-1.05
1992-93	6.2	5.8
1993-94	4.1	4.1
1994-95	5.1	5.0
1995-96	-1.1	-0.9
1996-97	10.1	9.6
1997-98	-2.8	-2.4
1998-99	6.9	6.2
1999-00	1.0	1.3
2000-01	-0.6	-0.4
Annual Average (1991-92 to 2000-01	2.70	2.68
Total Negative growth	-6.4	-5.2

Source: Computed from the Economic Survey, Government of India

III Trends in Foodgrain Sector and Food Availability

Similar trends are observed in the growth rate of foodgrains also. Table-4 shows the trends in the growth of foodgrains production in the country. During the period 1981-82 to 1990-91 the overall foodgrain growth rate was 3.46 percent with growth rates of rice and wheat at 3.99 and 4.57 percent respectively.

Table 4: Trends in Growth Rates of Foodgrains Production

Year	Rice	Wheat	Coarse cereals	Total Cereals	Pulses	Total Foodgrains
1981-82	-0.71	3.14	7.13	2.38	8.28	2.86
1982-83	-11.51	14.26	-10.74	-3.39	3.04	-2.84
1983-84	27.55	6.29	22.16	18.55	8.68	17.65
1984-85	-2.93	-3.10	-8.05	-4.24	-7.21	-4.49
1985-86	9.41	6.76	-15.94	2.62	11.71	3.37
1986-87	-5.12	-5.80	2.40	-3.92	-12.35	-4.67
1987-88	-6.11	4.17	-1.75	-1.76	-6.40	-2.14
1988-89	23.97	17.20	19.39	20.62	26.37	21.07
1989-90	4.37	-7.87	10.45	1.35	-7.15	0.66
1990-91	0.98	10.61	-5.93	2.50	10.89	3.13
1991-92	0.52	1.00	-20.52	-3.56	-15.71	-4.54
1992-93	-2.44	2.73	40.78	6.59	6.66	6.59
1993-94	10.21	4.60	-15.80	2.57	3.82	2.66
1994-95	1.88	9.91	-3.02	3.81	5.48	3.93
1995-96	-5.90	-5.58	-2.84	-5.27	-12.32	-5.79
1996-97	6.17	11.67	17.50	10.16	15.76	10.54
1997-98	0.99	-4.33	-10.88	-3.19	-8.98	-3.60
1998-99	4.29	7.45	3.06	5.25	14.96	5.90
1999-00	4.18	7.13	-3.16	4.08	-10.06	3.04
2000-01	-5.24	-8.76	2.44	-5.42	-17.45	-6.19
2001-02	9.84	4.43	7.37	7.40	20.78	8.15
Average Annual Growth Rates						
1991-92 to 2001-2002	2.23	2.75	1.36	2.04	0.30	1.89

Source: Same as for Table 3

Due to declining growth rate in food grain production, the actual availability of per capita foodgarins has also declined over the years, especially after 1996 (Patnaik, 2004) For the first time after independence, the decadal growth rate of foodgrains has been lower than the population growth rate in India (Table-5).

Logically with the opening of the economy, the fluctuations in per capita foodgrains should have been avoided (if the arguments given behind the signing of the WTO are to hold true). The situation however has been totally reverse. While, on the one hand the per capita availability has been declining, the export of foodgarins touched peak during some of the worst agricultural years, 2002 in particular (Patnaik).

Many plausible reasons have been given for the decline in the agricultural growth rate and also the decline in the per capita foodgrain availability and consumption. While government has failed to recognize the declining foodgrain absorption in the rural areas and justifies the slowing down of agricultural growth rates on the grounds of vagaries of nature, the critics ascribe to these failures directly to trade liberalization in agriculture sector following the signing of WTO agreement (Patnaik). The reasons for the declining foodgrain absorption, as evident from the declining foodgrain off-take from PDS (Jena, 2004), is usually ascribed to declining purchasing power among the rural masses.

Table 5: Annual Per Capita Foodgrains output and Availability in the Nineties

(Three Year Average)							
Three-yr. Period	**Average**	**Net Output per Head**		**Net**	**Avail-ability**	**per Head**	
Ending in	**Population**	**Cereals**	**Foodgrains**	**Cereals**	**Pulses**	**Foodgrains**	
	Million	**kg**		**kg**	**kg**	**kg/ year**	**Gms day**
1991-92	850.7	163.43	178.77	162.8	14.2	177	485
1994-95	901.02	166.74	181.59	160.8	13.5	174.3	478
1997-98	953.07	162.98	176.81	161.6	12.6	174.2	477
2000-01	1008.14	164.84	177.71	151.7	11.5	163.2	447

Individual year								
2000-01	1027.03	157.79	167.43	141.42	9.64	151.06	414
2001-02	1046.44	165.4	177.01	146.76	11.61	158.37	434
2002-03*	1066.22	140.54	150.09	148.14	9.55	157.69	427
Average of the Years 2000-01 & 2001-02*	1036.74	161.63	173.3	144.51	10.64	155.15	425
Change in Per Capita Availability of Foodgrains, %							
Triennium ending 1991-92 to Triennium ending 1997-98						-1.6	
Triennium ending 1997-98 to biennium ending 2001-02						-10.9	
Total Change, 1991-92 to 2001-02						-12.3	

Source: Patnaik (op.cit.)

(C)

Observations and Issues:

We may now briefly summarize the discussions presented in the preceding sections. During the past 10 years India has undertaken several policy reforms to create an enabling environment for opening up of the economy and liberalization of agricultural trade, as a part of broad economic reforms programme initiated during the nineties, and to meet it's commitments as a WTO member country. The formation of World Trade Organization as a multilateral trade regulatory authority in 1995 to catalyze the philosophy of free trade has raised serious question marks on the concept and philosophy of free trade itself. Theoretically the principal mandate of the WTO is to promote an open, equitable and non-discriminatory multilateral trading system based on the principle of comparative advantage. However, implementation of WTO framework and its agreements, have led to serious problems resulting into non agreement on a number of issues as they seem to follow the principle of discrimination and unequal exchange. The failure of recent round of Cancun summit to arrive at a consensus with regard to Agreement on Agriculture is a clear pointer. The three pillars of AoA, i.e., increasing market access, reducing domestic support (specially the subsidies on farm inputs) and curtailing export promotion measures (like export subsidies) under the WTO framework of negotiations, as of now, do not warrant for any argument and concessions to the developing countries with regard to the issues of self sufficiency, food security and poverty reduction through agriculture. This has led to serious policy dilemma and dichotomy in the country as regards to the objectives of a sound agriculture policy.

While India seems to have acted under pressure and presumptions to highlight the need for reducing the domestic support, crop diversification and export enhancing measures, the objective situation has always demanded a food security approach as nearly eighty percent of the farmers belong to small and marginal category, hardly having any competitive strength and economic security. The growth rates in agriculture sector have already started declining and the foodgrain situation appears to be worse toady despite the surplus production. The dichotomy becomes more acute when the issue of agriculture is viewed in relation to poverty reduction, which not only requires government support to the farm sector but also direct intervention calling for asset redistribution and land reforms and extending protection to the farmers and agricultural labourers. The issues of asset redistribution and land reforms have totally been relegated in to background ever since the export frenzy approach to agriculture growth took primacy. While the export frenzy approach is based on large farm economics, the issues of food security, environmental sustainability and preservation of bio diversity require a completely decentralized approach in which the local institutions and the class of small and marginal farmers may have the adequate and effective participation. It is apparent that under Indian conditions the issue of agriculture growth, with or without WTO, can not be separated from the issue of agrarian relations which determine the level of food security, environmental sustainability and poverty. The issue of environmental sustainability in Indian context can not be separated from the issue of food security as the loss of bio-diversity has been one of the prime causes for food insecurity. The loss of biodiversity has made small and marginal farmers more vulnerable after globalization despite the fact that this group had been the traditional protector of plant varieties in the country.

Neglected Issues and Policy Options:

As we have seen above, the down fall of agricultural situation in India is very obvious. While WTO and liberalization could have played a role in this downfall, the internal economic environment has equally been responsible for this situation. Continued ignorance of serious policy issues within the country has resulted into a situation of non competitiveness of Indian products in the world market and food crisis in the internal economy. Keeping above factors in view, some of the areas of critical importance for policy making can be listed as below-

1) **Targeted support to the Small and Marginal farmers**. With nearly eighty percent of the farmers belonging to the category of small and marginal farmers with more than one third of cultivated area under their possession, no policy reforms can succeed without targeting them. Over the years the benefits of subsidies to boost agricultural production under the policy framework of New Agricultural Technology (NAT) and Foodgrain Self Sufficiency, has mainly gone to the large and medium farmers. Majority of them have now become asset surplus operators. The fruits of the domestic support system yet to reach the small and marginal farmers. The basic issue is "how to ensure it?"
2) **Role of Minimum Support Price in Crop Diversification**. With launching of the new agricultural technology, the minimum support price system was devised as a tool to encourage the farmers for adoption of HYV seeds and orient the production system in favour of foodgrains. Today the NAT itself is under attack due to its adverse ecological impact. Some of the states have started the practice of announcing the MSP for a wide range of products. But in the absence of any buy back system on part of the states, the policy has been totally ineffective. The main issue, therefore, is "How the MSP can be used as an effective policy instrument to reorient the cropping pattern

suitable to the ecological conditions and also to serve the interests of small and marginal farmers"?

3) **Land use pattern and food security:** Food security has emerged as one of the prime concerns in the post globalization period. Indian farming sector is one of the largest private sector occupation based on the principle of perfect competition. This excludes the possibilities of targeting the production systems by direct intervention. Hence the issue is: "what are possibilities and scope for local level land use planning by participation of the land owners to sub-serve the needs of the local level food security and maintenance of bio-diversity"?

4) **Restructuring Land reforms:** The onset of globalization has relegated the issue of land reforms totally in the background. Worse still, it has had the impact of even altering the concept of land reforms and most of the talks presently are confined to ensuring the interests of the capitalistic farming serving the interest of the few. The issues involved in land reforms area are much more diverse and complex than mere talk of contract farming. The need for restructuring land reforms policy is needed in respect of following

- land ownership and contract farming
- Enhancing labour absorption capacity of the farm sector.
- Increasing the access to land, especially of Common Property resources to the poor,
- Environmental challenges and land and water management.
- Institutional support system for viability and sustainability of small and Marginal farmers.
- Institutional mechanism for effective participation of small and marginal farmers.
- Promoting Integrated Farming System.

Tackling these issues requires no international negotiations, the agriculture situation and therefore there is urgent need to focus on these aspects to improve agriculture situation in India. Some of the states which tried to tackle to these issues have shown remarkable performance in this regard. The successful implementation of watershed programme and irrigation development in states like Rajasthan, Andhra Pradesh, Karnataka, Gujarat, etc., show very clearly that environmental hazards and the problems of low productivity and growth can be resolved successfully and simultaneous.

References:

Choudhury, RC and R P Singh' 2000; 'Rural Prosperity and Agricultural Policies and Strategies', NIRD

CSO; 'National Accounts Statistics', Government of India, Different Issues

Government of India; Economic Survey, Various issues

Patnaik, U. 2004; 'The republic of Hunger', Public Lecture on the Occasion of the birth day of Safdar Hashmi, 10th April, 2004

Reserve Bank of India, 2003; 'Report on Currency and Finance'

Thapliyal, B K, 2002; 'WTO and Indian Agriculture: Challenges and Opportunities' Mimeograph, NIRD

* * *

3

DAIRY AND LIVESTOCK DEVELOPMENT IN MEGHALAYA

C.S.Choubey[1] AND **B.K.Mishra**[2]
[1] Department of A H & Dairying, K.A.P.G. College, Allahabad, U.P,
[2] Department of RDAP, North-Eastern Hill University, Tura Campus,Tura-794002, Meghalaya

ABSTRACT

Livestock sector plays a significant role in the Indian economy, particularly for the welfare of rural population of India. Of the total household in the rural area, about 73 percent own some form of livestock. More importantly, small and marginal farmers account for three quarters of these households. The focus of the poor is on small animals. The prolificacy of goat, pig and poultry are the influencing factors. The returns are quick; losses, if any, are recovered soon. The poor can afford it. The multiple species animal husbandry system is also environment friendly. This sector plays an important and vital role in providing nutritive food, rich in animal protein to the general public and in supplementing family incomes and generating gainful employment in the rural sector, particularly among the landless, small, marginal farmers and women. Income from livestock production accounts for significant percentage of total farm household's income in different states. Thus, an increase in demand for livestock products can be a major factor in raising the income and living standards of the rural household. With its large livestock population, India also has vast potential for meeting the growing needs of teeming millions, particularly in respect of livestock products such as milk, eggs, meat and wool. Under-privileged livestock producers face a number of constraints which affect the productivity of animals. Their access to modern livestock services, especially veterinary services are poor. Their access to the marketing support is also poor. The prices received by them for products are low. For example, majority of milk producers get only 50 percent of the price paid by consumer. Credit support for purchase of animals and its maintenance is not easily available to the small holders. Goat, pigs and backyard poultry are most commonly kept by the underprivileged and these get very little development/research support. Improved animals that would 'niche well' with the systems of the underprivileged and adverse agro-ecological conditions in which majority of them live, particularly the ecologically fragile regions, are not available. There are hardly any programmes to assess their needs and produce or make available such animals. Feed, fodder availability is a major constraint and conventional approach to improve the situation does not work with the underprivileged. Being a hilly state most farmers practice shifting cultivation and the food production is perennially deficient of its growing demand. This in turn affects the supply of the livestock feed ingredients. Thus, the livestock production stands as backyard venture rather than an economically viable enterprise. On account of scarcity of feed, the farmers are unable to take large scale livestock production. Mineral resources generate sizable revenue to the government, which is also not reinvested in other sectors. But the externality of mining activity in terms of man-made problems of water scarcity adversely affects the crops, people and livestock. The household stock of livestock population is extremely low and the houses are scattered making vaccination of the common disease extremely difficult. Lack of proper road and communication made it difficult to reach the people in the time of crisis, which aggravate the situation.

Keywords: Livestock, dairy, farming, constraints, contribution.

Introduction:

Meghalaya (literally the abode of clouds) attained statehood on 21 January 1972. The state was carved out of erstwhile Assam and came into existence as full-fledged state. The state is land-locked territory of lovely hills and abounding sylvan beauty and natural resources. It has a temperate climate and rainfall is high. North, south and western border are comparatively warmers in winter due to lower altitude. It has four distinct seasons, spring time (March-April), summer or rainy season (May-September), Autumn (October-November) and Winter (December-February). The total area of the state is 22429 square kilometres, most of which is hills with a population of 1774778 (1991). The density of population is 79 with 81.4 per cent in rural areas and 18.6 per cent in urban areas. The sex ratio is 955 female. Predominantly Meghalaya is inhabited by tribal population consisting of 85.5 per cent S/T and 0.51 per cent S/C and 13.9 per cent general population. The Khasis, Jaintias and Garos are the main tribal group. There are seven administrative divisions/districts in the state. About 49.1 per cent is the literacy rate ranging from 81.74 per cent urban and 44.85 per cent rural literates. The main workers are 40.3 per cent of which 55.3 per cent cultivators and 12.5 per cent agricultural workers. The participation of womenfolk in the economic activities is predominantly in the state. Per capita income is about Rs 5519 and the population below poverty line is 33.87 per cent (1999-2000). The population of the state is mostly meat eaters and the consumption of milk is negligible. It indicates the need for requirement of livestock product in the state.

The mode of cultivation is mainly hill terrace farming, but the productivity is low. The agricultural economy of the state is characterised by subsistence nature. The livestock population is dominated by cattle with 6.98 lakh indigenous and 16.8 thousand crossbred cattle. Pig population is also substantial at 2.94 lakh, the next is 1.96 lakh goats and the poultry; desi 1.68 lakh and only 4.5 thousand improved poultry. There are a few number of ducks, drakes and ducklings about 45.4 thousand. The veterinary service facilities in the state are not satisfactory.

Contribution of Livestock and Dairy Production for Balanced Ecological Cycle:

Though pastoral practices still exist in the state, and there are pastoral communities, predominant production mode is one of the mixed farming in which agriculture and animal production are interdependent. Agriculture provides energy to livestock in the form of by-product feed, and the livestock return a part of that energy in the form of draught power and manure besides producing milk, meat and other animal products. Not more than 5 – 6 per cent of the area under cultivation is allocated to fodder crop production to supplement the by-product feed with green fodder.

The positive environmental contribution of livestock and dairy production systems are due to:

I) **Recycling agricultural by-products as animal feed**: The major land savings occur because the alternative is to produce equivalent amount of green fodder by allocating require land area.

II) **Use of dung as manure**: Since manure substitutes for chemical fertilizers, besides providing plant nutrients (NPK), it protects soil-born micro organisms, and prevents green house gasses emission that would otherwise occur in the manufacture, transport and distribution of equivalent amounts of chemical fertilizers.

III) **Use of dry dung cake as domestic fuels**: The land savings occurs because the alternative is to plant and harvest firewood, trees, and supply equivalent amount of dry firewood. Since there is gestation lag between planting and harvesting of firewood, it requires much land than the area required to be harvested each year.

IV) **Use of working animals in agricultural operations and also for rural transport:** The working animals substitute for tractor and other agricultural machinery run on fossil fuel. The working animal stock prevents green house gasses emission, in particular CO2 emission that would otherwise occur due to burning of fossil fuel in running the substitute number of tractors. Incidentally, whereas fossil fuel is a non-renewable resource, the working animal stock is renewable.

Scenario of Livestock Production in Meghalaya:

The state is perennially deficient of livestock products. The total short fall is 7.3 thousand tons of beef (total requirement 28.8 thousand tonnes), 900 tonnes of pork (total consumption 8500 tonnes), and 32 million eggs out of total requirement of 1116.7 million. The consumption of mutton and chicken is negligible. In order to promote the sector to meet the growing demand for livestock product, the trained manpower is also scarce.

Piggery Development:

Pig is one of the most important species among the small animal in the State. Pig rearing is almost a common occupation with the tribal people of the state both in the plain areas of Garo hills and in the Hill region of Khasi and Jaintia hills. The suitability of the climate, fondness to pigs among the major section of the State population and availability of good market for pork promise wide possibilities of development of the pig industry in the State. The main drawback in this sphere as in the case of other species of livestock is the inferior equality of stock. Most of the pigs reared are being of non-descript local breeds. About 88% of the total villages reared non-descript local breeds and only 12% reared Cross Bred pig. *Rs. 78.55 lakhs* was provided under a Scheme during the year 2002-03.

With a view to upgrading local stock, the Department of Animal Husbandry and Veterinary has taken up schemes to introduce improved breeds of pigs such as Yorkshire saddle back, Hampshire etc. These improved breeds of pigs are kept at Government pig farm established in every district for the purpose of demonstration as well as for producing pedigree stock for distribution to the interested breeders through Livestock breeding programmes. A Regional Pig Breeding Farm was established at Kyrdemkulai during the 7th Plan period in the year 1986-87 with financial assistance from N.E.C. The farm, which is now under the State plan, is maintaining Hampshire saddle back, Yorkshire pigs for scientific breeding purpose. The farm is producing piglets and supplying to the entire District farm for raising and multiplication and supplying of these piglets to farmers. Under Piggery Development there were 10 (ten) pig farms including the Regional Pig Breeding Farm Kyrdemkulai established in the state.

Another Pig Farm has been set up at Sohra which is yet to function. Construction of buildings has been completed.

(a) Achievement in Government Farm:

S. No.	Name of the Farm / Location.	Total Stock Reared	Breeding Stock		Total Young Stock / Growing Stock / Piggling Stocks	Piglet Produced	Pig Sold
			Boar	Sows			
1	Pig Farm, Mawryngkneng.	23	1	11	11	81	78
2	Pig Farm, Rongkhon.	73	4	14	55	154	134
3	Pig Farm, Thadlaskein.	222	3	44	175	480	456
4	Pig Farm, Nongstoin.	40	4	8	28	65	42
5	Pig Farm, Rongjeng.	31	1	10	20	34	11
6	Pig Farm, Mairang.	41	3	14	24	97	79
7	Pig Farm, Baghmara.	47	2	6	39	66	4
8	Pig Farm, Dalu.	55	2	13	40	100	94
9	Pig Farm, Pynursla.	80	7	14	59	110	66
10	Regional Pig Breeding Farm, Kyrdemkulai.	271	15	55	191	1013	789
	Total	**883**	**42**	**189**	**642**	**2200**	**1753**

(b) Distribution of Piggery unit (General) @ 50 % subsidy was made to the selected beneficiaries; the total unit distributed was *56* Units with *10 (ten)* Nos. of pigs per unit. The objective of this programme is to enable the identified beneficiaries to take up Piggery as subsidiary occupation to supplement their income.

(c) Employment Generation to Educated Unemployed youth continued during the year under report and *37*units were distributed to selected beneficiaries @ 50% subsidy with *5 (five)* gilts and 1 (one) boar per unit of Improved breed.

(d) Scheme for Piggery production programme under SLPP @50% subsidy to individual beneficiaries for taking up of Piggery production programme for SF/MF continued as usual. The unit costs include *3 (three)* gilts and *1 (one)* boar of Improved breed including construction of pigsty, and concentrate feed, equipment's etc.

Cattle Development:

This is an integrated programme of Cattle Development including breeding and management practices. The main objective of the scheme is to increase Milk production in the state by adopting improved breeding practices through artificial insemination programmes. Four Cattle breeding farms were established in the State with Exotic breeds like Jersey and Holstein-Friesian for maintenance of superior germplasm and enhancement of Milk production. During the year 2002-03 the total plan allocation for the cattle development programmes was *Rs. 68.87* lakh as per revised Outlay.

Cattle and Buffalo Breeding Farms:

Sl. No.	Name of the Farm	PARTICULARS OF STOCK MAINTAINED (Nos.)									
			Milch Cows					Calves Reared			
		Total Stock Reared	In Milk	Dry	Total Milch Cows	Young Bull	Heifers	Male	Female	Calves produced during the year	Milk Produc-tion in '000' Litres
1	Indo Danish Project, Upper Shillong.	133	30	33	63	-	22	18	30	37	136.1

2	Regional Cross Bred Cattle Breeding Farm, Kyrdemkulai.	85	22	15	37	1	23	11	13	22	43.2
3	Livestock Farm, Rongkhon.	78	18	13	31	8	26	2	11	16	65.7
4	Cattle Farm, Khliehtyrshi.	40	10	12	22	-	7	1	10	9	42.1
	Total Cattle :	336	80	73	153	9	78	32	64	84	287.1
5	Buffalo Farm, Songsak	35	6	7	13	7	8	5	2	8	5.8

Under Intensive Cattle Development Project, Shillong and Tura: Nineteen thousand improved calves were produced during the year 2002-03 in different areas of the State through Artificial Insemination programmes implemented under this Scheme. In the same year twenty four thousand artificial inseminations were carried out in the State to produce improved calves for augmenting Milk production.

Assistant to SF/MF/AL for rearing of cross bred heifers: The main objective of the scheme is to enable the identified beneficiaries to raise up quality crossbred Heifers. The scheme is also to give inputs as part of overall programmes for Dairy Cattle Development in the State. But the Scheme could not be implemented during the year due to non receive of sanction from the Government.

Distribution of Bull/Calves/Cows Grant-in-Aid : The objectives of the Scheme is for rearing of cross bred Heifers /Calves for Dairy farming to provide incentive to farmers to take up Dairy farming to generate self employment to the people and increase in production of Milk. During the year 2002-03 *9* units with 2 (two) crossbred cows per unit were distributed accordingly to selected beneficiaries. The assistance was *50 %* subsidy.

Scheme for educated un-employed youth: Scheme for educated Un-employed youth is to generate employment through Dairy farming with 50% subsidy. Scheme not sanctioned during the year 2002-03.

Sheep and Goat Development: The objective of the present Scheme is to improve the local goats by introducing superior germplasm. The Goat Farm Nongshillong, reared 4 Bucks and 47 Doe, with a production of 52 Kids. The Sheep and Goat Farm Saitsama continued its function during the year with 1 Bucks and 5 Doe with production of 10 Kids. As regard to rearing of Sheep, the farm reared 5 Ram and 9 Ewes with a production of 6 Lambs.

Subsidy Schemes for distribution Goatery unit:The Scheme was implemented during the year 2002-03 with 20 Units (50 % Subsidy). The beneficiaries were benefited with 5 local Goats and one Improved Buck per unit. 20 Unit was sanctioned in the same year.

Rabbit Farm Nongpiur was established with an objective to produce the breeding stocks for supplying to farmers to increase meat production. The farm maintained 110 Nos. of parent stock with a production of 680 Nos. of Rabbit.

Status of Livestock Diseases:

Coccidiosis and Ranikhet diseases are the most epidemic diseases with 13626 attacks and 522 deaths, and 5895 attacks and 456 deaths respectively in 1999-2000. The next important disease is chronic respiratory and swine fever. The foot-and-mouth disease, fowl fox and rabies are other important diseases.

Problems and constraints:

Being a hilly state most farmers practice shifting cultivation and the food production is perennially deficient of its growing demand. This in turn affects the supply of the livestock feed ingredients. Thus, the livestock production stands as backyard venture rather than an economically viable enterprise. On account of scarcity of feed, the farmers are unable to take large, scale livestock production.

Mineral resources generate sizable revenue to the government, which is also not reinvested in other sectors. But the externality of mining activity in terms of man-made problems of water scarcity adversely affects the crops and people and livestock.

The household stock of livestock population is extremely low and the houses are scattered making vaccination of the common disease extremely difficult. Lack of proper road and communication made it difficult to reach the people in the time of crisis, which aggravate the situation

Dairy development in Meghalaya:

The objective of the latest scheme in the state is to make overall development in the Dairy sector. Under the Scheme, supply of wholesome Milk to the consumers after procuring the Milk from the Dairy Co-operative Societies / Farmers and pasteurising the same in the Departmental dairy Plants. Dairy farming subsidy schemes are also being implemented for enhancement of Milk Production.

Number of milk plants and milk Centres with capacity- District-wise.

Sl. No.	District	Dairy Plant		Milk Chilling Centres	
		Centres (Nos.)	Capacity '000' lt.	Centres (Nos.)	Capacity '000' lt.
1	East Khasi Hills.	1	10.0	-	-
2	Ri Bhoi.	-	-	-	-
3	West Khasi Hills.	-	-	1	2.0
4	Jaintia Hills.	1	2.0	1	2.0
5	East Garo Hills	-	-	1	2.0
6	West Garo Hills	1	2.0	-	-
7	South Garo Hills.	-	-	-	-
	State :	**3**	**14.0**	**3**	**6.0**

Achievement during the year2003-04 under report is shown below:

Sl. No.	Name of the Scheme	Quantity of Milk Procured (Tonnes)	Quantity of milk distributed (Tonnes)	Quantity of Milk Products / Produced		
				Cream Kg.	Butter Kg.	Ghee Kg.
1	Central Dairy, Shillong.	751	735	597	82.5	368
2	Central Dairy, Tura.	134	129	81	48	34
3	Creamery & Ghee making Centre, Tura.	14	5	567	185	143
4	Rural Dairy Extension Centre, Jowai.	69	53	146	98	-
5	Chilling Centre, Nongstoin.	9.5	9.3	-	-	-
6	Chilling Centre,Gangdubi	8.5	7.9	-	-	-
	Total:	**986**	**939.2**	**1391**	**413.5**	**545**

However the total estimation of milk production in the state (Meghalaya) from the year 1997 to 2009 is shown in the table below:

YEAR	PRODUCTION IN(000) TONES
1997-98	59
1998-99	61
1999-00	62
2000-01	64
2001-02	66
2002-03	68
2003-04	69
2004-05	71
2005-06	73
2006-07	75
2007-08	77
2008-09	77

Source: Department of Animal Husbandry, Dairying & Fisheries, Ministry of Agriculture, G0I

The above table shows that the production of milk in the state Meghalaya has been increasing year after year, from the year 1997- 2009.

Conclusion:

Due to its peculiar Topography, economy of the people of Meghalaya depend more on the sectors like Animal Husbandry, Horticulture etc. as compared to traditional Agriculture. To bring about a developmental change in livestock and dairy sector, number of programs and infrastructure development are being taken up.

References:

http://www.shvoong.com/exact-sciences/1885501-dairy-livestock-farming-meghalaya-potential/

http://www.shvoong.com/exact-sciences/agronomy-agriculture/1891255-livestock-dairy-farming-meghalaya-policies/

http://www.ncap.res.in/upload_files/workshop/wsp10/html/chapter7.htm

http://www.nddb.org/statistics/milkproduction_states.html

http://meghavt.gov.in/repannual02 - 03.htm

* * *

4

SUSTAINABLE RURAL POULTRY PRODUCTION IN NORTH EAST REGION: APPROACH AND STRATEGY

S. Malik[1], D. K. Pandey[2] and M. Datta[1]
[1] ICAR Research Complex for NEH Region, Tripura Centre, Lembucherra – 799 210, Tripura West, [2] College of Fisheries, CAU, Lembucherra – 799 210, Tripura West.

ABSTRACT

Rural poultry production is one of the major components of agriculture. Majority of birds in N. E. region are nondescript/deshi, in which growth and production is very low. Due to this reason, the per capita availability of eggs and meat in N. E. region is very low, which is far below than the recommended allowance i.e. 180 eggs per annum as suggested by the NAC of ICMR. A strategy is given in order to fulfill this deficiency gap and to make the region self sufficient in poultry production. First; there should be location specific variety for this region, which would be developed using local germ plasm of this region. Second; there is a need to develop cheaper balance ration using locally available ingredients. Third; supply of chicks on regular basis to the farmers of this region is most difficult task. Therefore, the Animal Husbandry Department of the concerned state Govt. should estimate the requirements according to actual need of the farmers in the state and should make the arrangement for supply of chicks. Fourth; training to the farmers is also required for the scientific management of the poultry production. It would help in increasing poultry production and reducing the diseases.

Introduction:

Rural poultry production is one of the major components of agriculture in North East region. Majority of birds of this region are deshi, in which growth and production is very low (60-70 eggs per bird per annum; 2.5 kg. and 1.5 kg. body weight per bird in males and females, respectively). Due to this reason, the per capita availability of eggs and meat of this region is very low. Moreover, the agriculture of N.E. region is characterized by monocropping system. The system of cropping is not suitable for providing round the year farm income. Therefore, most of farmers of this region rearing pig, goat and poultry in the back yard of their house to support their livelihood throughout the year. The N.E. region is most suitable for rural poultry production compared to other species of the animals. Rural poultry production has an appreciable advantage of being easy to raise, low input, higher growth and reproduction. Rural poultry farming utilizing surplus coarse grains, kitchen waste and greenery and converting them into protein rich human food and valuable manure. The manure obtained from poultry having more essential nutrients (three times the cattle manure) Nitrogen, Phosphorus, Potash, Magnesium, Sodium and Calcium plus trace minerals.

The average per capita availability of eggs in north East region is only 22.97, which is far below than recommended allowance viz. 180 eggs per capita per year (@ half egg per day per person) as suggested by Nutritional Advisory Committee of ICMR. Therefore, the short fall in egg production in NE region is 62,058.80 lakh nos. (2003-04). Thus it is warranted that production performance of rural poultry should be improved, which is essential for food security, employment generation as well as for poverty alleviation in rural areas.

Present Status and need of Improved Rural Poultry Germplasm for Sustainable Rural Poultry Farming:

The rural poultry production can play a greater role to fulfill the requirement of egg and meat production in North East region. It is difficult to establish commercial layer farms in N.E. region due to very high cost of feed and socio political problems.

On an average the egg production of good commercial layer is 300 eggs in a laying year. So the per day average egg production is 0.82 egg and per day feed consumption is approx. 110 gm. Therefore, the feed consumption per egg is 134 gm. that cost Rs.2.68 (@ Rs.20/- kg.). Even this is the 70% cost of one egg as the feed cost is approx. 70 % of the total cost of the farm. So the total cost per egg is Rs. 3.82, which is much higher than the cost per egg in the market i.e. Rs. 3.50, which procured from the other states. So it is very difficult to run commercial layer farm successfully in North East region. However, cheaper feed if prepared using locally available ingredients may help in developing small and medium scale rural poultry units in the region. Hence, in the prevailing conditions, rural poultry production would be the only option to fulfill the requirement of egg and meat production. So that the egg requirement of N.E. states could be met from rural poultry production using improved germplasm.

The per capita per year egg availability in North East is only 23 (table -1) therefore, every state of north east procuring eggs in larger nos. from other states. However, if 75 % of the total deshi birds, which have the production potential less than 100 eggs in a laying year is replaced by improved stock (laying more than 150 eggs in a year), than more than 6,178.00 lakh eggs could be produced without much extra effort (table-2). By this way, at least procurement of eggs from other states may be avoided (Malik and Singh, 2009). To make the region self sufficient approx. 18.79 lakh farmers are required to rear at least 10 birds of improved stock in their backyard (table-2). For this purpose, the breeding farms of this region have to maintain approx. 2.95 lakhs females of improved parent stock along with approx. 15 % breeder males (table – 3).

Table 1: Status and requirement of egg production of Tripura, North East and India (2003 – 04).

	Total egg production. (lakh nos.)	Per capita availability (nos./year)	Total requirement of eggs prod. @ 180 eggs/ year. (lakh nos.)	Present prod. Short fall. (lakh nos.)	70% requirement to be fulfilled by the chicken (lakh nos.)**
Tripura	1005.00	31*	5,940.00	4,935.00	3,455.00
N.E.	9,079.00	23*	71,137.80	62,058.80	43,441.16
India	4,04,031.00	40*	18,18,140.40	14,14,109.40	9,89,876.58

* Including Duck egg. ** Remaining 30 % to be fulfilled by duck.

Table 2: Planning to fulfill the egg production requirement by using improved rural poultry germplasm.

(in lakh nos.)

	Population of deshi layer (2003)	Present egg production by deshi birds (2003-04)	25% deshi birds with prod. Potential 100 eggs or more /year remained and 75 % replacement by improved germplasm (@ 150 eggs/year).	Total egg production (25% Deshi + 75% proposed improved stock + Existing improved stock)	Balance of egg production from shortfall.	No of farmers required to rear @ 10 birds of improved germplasm.
Tripura	7.49	643.00 (Av. 85.85)	1029.88 (+386.88)	1391.88	2063.12	1.375
N.E.	78.06	4,555.00 (Av. 58.35)	10,733.00 (+ 6,178.00)	15,257.00	28,184.16	18.79
India	807.58	83,896.00 (Av. 103.88)	1,21,137.00 (+37,241.00)*	4,41,272.00	5,48,604.58	365.74

*In case of national level 25 % remained at higher production of 150 eggs / year because at national level the average is already 103.88 eggs / year so the remaining 75% of poor producer may be replaced.

Table 3: Requirement of chicks and Parent stock to fulfill the requirement of egg production.

(in lakh nos.)

	No. of commercial females required to be rear by the farmers	No. of chicks required for maintaining these females (50% Sex ratio and 10% mortality)	No. of fertile eggs required for the chicks (Av. 70% hatchability on TES)	Requirement of no. females of parent stock (Av. Egg production 200 eggs in a laying year)
Tripura	13.75	30.25	43.21	0.216
N.E.	187.90	413.38	590.54	2.95
India	3657.40	8046.28	11494.69	57.47

Improved rural poultry germplasm developed by different Institutes/ organization:

The University of Agriculture Sciences, Bangalore first developed dual purpose improved coloured backyard bird named Giriraja in 1989 (Viroji Rao, 2006). Later on several varieties viz. Vanaraja, Krishibro, Gramapriya, Krishilayer and krishipriya by project Directorate on Poultry, Hyderabad; CARI Nirbheek, CARI Shyama, Hitcari and UpCari by Central Avian Research Institute, Iztanagar (U.P.); Nandanum by Tamilnadu University of Veterinary and Animal Sciences, Chennai, Swarandhara by Veterinary, Animal and Fishery University, Habbal (Bangalore) and Kroiler by Keggs farms etc. were released. Most of these varieties developed by crossing local indigenous germplasm with improved germplasm. The crossbred progeny improved body weight and egg production compared to their parental indigenous germplasm. The main problems observed in dual purpose stock viz. Giriraja, Vanaraja, Krishibro, Nandanum and Swarandhara etc. are their non flying capacity due to heavy body weights so prone to predation, require more supplement feed and non broody so difficulty in multiplication. Therefore, subsequently some egg type backyard birds were developed viz. Gramapriya and krishipriya by PDP, Hyderabad; Krishna J. by JNKVV, Jabalpur; CARI Nirbheek and CARI Shyama by CARI Izatnagar (U.P.);

Rajasri by Sri Venkateswara Veterinary University, Hyderabad and kalinga Brown by C.P.B.F., Bhubaneswar, which have moderate body size with good egg production potential.

Constraints in the development of backyard poultry in N.E. region:

1. Lack of location specific variety for this region.
2. Insufficient infrastructure and parent stock to fulfill the requirement of good chicks.
3. Unavailability of low cost feed to use as minimum supplement of feed.
4. Problems of vaccination and disease control.
5. Lack of technical knowledge of poultry farming in rural people.
6. Negligence and attitude of rural people towards poultry.

Evaluation of rural poultry germplasm - ICAR Tripura Centre experience:

ICAR sanctioned a Centre of All India Coordinated Research Project on Poultry Breeding (Rural Poultry Production) at ICAR Tripura Centre in the year 2000-01. During the period, the performance of seven germplasm namely, Vanaraja, Giriraja, Gramapriya, Krishna J., CARI Nirbheek, CARI Shyama and Tripura native germplasm (black) were evaluated at the Institute farm and farmers house under this project. Vanaraja and Gramapriya germplasm were procured from Project Directorate on Poultry, Hyderabad; Giriraja and Krishna J germplasm were procured from University of Agricultural Sciences, Bangalore and JNKVV, Jabalpur respectively; CARI Nirbheek and CARI Shyama were procured from CARI Izatnagar (U.P.).

Out of the total number of chicks hatched, half of the survived chicks at 6 –8 weeks of age were distributed to farmers. Remaining fifty per cent are being maintained at Institute farm. Growth rate and production performance of the birds were estimated at farm as well as under backyard system at farmers' field.

The body weights at all the ages, fertility and hatchability were higher in Giriraja followed by Vanaraja than Gramapriya, CARI Nirbheek, Krishna J., CARI Shyama and Tripura native germplasm both at farm as well as farmers house (table.4 & 5). The feed consumption was higher in Giriraja followed by Vanaraja, and lowest in Tripura native germplasm. The higher body weight in Giriraja and Vanaraja than other stock is due to the genetic make up of the birds being produced by the crossing of broiler germplasm. The higher feed consumption may be attributed to better growth rate. Age at sexual maturity was lowest in Gramapriya than all other stock, therefore; the egg production upto 40 weeks of age was higher in Gramapriya at both the locations. The broodiness was not observed in all the improved germplasm but due to heavy body weight only mild flightiness was observed in Giriraja, Vanaraja and Gramapriya therefore the adult mortality due to predators was observed upto 40% at the house of some farmers. The adult mortality due to predators was observed lowest in case of Tripura native germplasm followed by CARI Nirbheek and CARI Shyama being lower body weight of the birds (Annual Reports, 2001-09).

It was observed that Gramapriya birds could thrive well on moderate supplementation in comparison to Vanaraja, Giriraja and Krishna J. under scavenging system of rearing. Farmers were happy with egg production as well as growth rate and they are ready to keep Gramapriya birds, especially females for longer period because of their better egg production. The Tripura native germplasm produced 124 eggs upto 72 weeks of age at the ICAR Tripura Centre farm but reported only 50-60 eggs per year at farmers field, the main reason for lower egg production at farmer's level is broodiness of the birds. The farmers should allow brooding, only when it is require and that may be one or two birds, other birds should not allow to sit on the eggs, eggs should be immediately collected after laying, by this way farmers can increase enough egg production in the native germplasm.

Table 4: Mean Performance of economic traits of different rural poultry germplasm at institute farm.

Traits	Vanaraja	Giriraja	Gram-apriya	Krishna J.	CARI Nirb-hic	CARI Shyama	Tripura native germplasm (black)
Fertility %	86.08	91.51	81.25	80.19	81.82	79.49	81.00
Hatchibility %							
(I) TES	74.95	82.35	64.71	75.94	80.82	49.86	67.00
(ii) FES	87.06	90.00	79.97	94.69	98.77	62.72	82.00
4Wk BW (g) M	377.25± 5.13	405.97± 16.13	312.73± 4.63	292.30± 11.53	258.88± 7.49	235.88 ± 9.47	162.66 ± 3.07
F	327.16± 4.35	362.23± 20.19	285.62± 3.76	264.92± 14.30	(pooled)	(pooled)	148.56 ± 2.15
8 Wk BW (g) M	1223.24± 10.15	1425.93± 10.83	751.47 ±10.13	797.44± 12.04	531.84 ±12.5	545.50 ± 17.97	355.49 ± 8.99
F	1113.19± 12.81	1353.28± 7.57	625.02 ± 12.76	592.86± 12.24	457.86± 9.72	414.54 ± 9.03	296.30 ± 5.18
12 Wk BW (g) M	1454.92± 15.04	1785.27± 17.59	1358.79 ± 17.19	1068.43± 30.10	932.36 ± 20.39	873.34 ± 22.70	620.79 ± 11.12
F	1297.27± 10.58	1650.96± 15.34	1164.36 ± 19.16	986.50± 28.20	769.86 ± 12.86	725.59 ± 27.09	504.83 ± 7.68
20 Wk BW (g) M	2581.64 ± 51.44	2782.94 ± 42.46	2107.65 ± 39.67	2000.94± 68.05	1801.82 ± 19.99	1594.07± 24.48	1186.09 ± 25.46
F	2055.94 ±33.15	2243.09 ± 56.41	1613.05 ± 43.97	1548.57± 42.24	1414.69 ± 17.56	1195.36 ± 21.18	1003.64 ± 13.99

40 Wk BW (g) M F	3721.25 ± 51.90 2710.36 ± 48.86	4274.67 ± 63.62 3258.63 ± 77.45	3213.82 ± 53.83 2331.78 ± 47.65	2820.12± 80.62 1604.11± 58.15	2666.60 ± 42.40 2090.76 ± 31.01	2.073.45 ± 45.27 1777.24 ± 35.38	1708.74 ± 39.55 1317.85 ± 24.80
Age at first egg (Days)	151.00	162.00	127.00	142.00	187.00	193.00	160.00
Egg number up to 40-weeks of age.	38.13	32.63	62.13	16.40	37.32	29.76	36.17 ± 1.54
Egg weight at 40 week of age (g).	61.53 ± 0.18	64.83 ± 0.28	59.25	51.32	53.78 ± 0.09	53.32 ± 0.38	38.18 ± 0.50
Cumulative feed consumption upto 6 weeks of age (g).	1998.74± 30.13	2376.80± 37.37	1165.00	-	981.25	969.06	762.35
Feed consumption per day between 19-20 weeks of age (g).	128.87 ± 11.62	142.99 ± 18.16	103.54± 10.36	-	99.16	97.42	88.30
Feed consumption per day between 39-40 weeks of age(g)	223.05 ± 14.13	254.50 ± 12.63	163.88± 9.23	-	160.22 ± 0.68	158.50 ± 0.80	89.34

Table 5: Mean Performance of different economic traits of six rural poultry germplasm at farmers' level.

Traits		Vanaraja	Giriraja	Gram-apriya	Krishna J.	CARI Nirbhic	CARI Shyama
8 Wk BW (g)	M	1006.21± 20.54	1221.42± 37.90	-	-	-	-
	F	921.52± 11.97	1052.79± 32.85				
12 Wk BW (g)	M	1321.43± 48.60	1593.69± 96.58	1084.22 ± 21.35	-	929.60 ± 17.65	813.02 ± 27.64
	F	1098.76± 25.50	1364.93± 78.15	928.15 ± 29.14		730.61 ± 15.03	594.53 ± 19.19
20 Wk BW (g)	M	2318.92 ±83.02	2661.76 ± 62.68	1699.36 ± 44.19	2115.86± 86.19	1641.70 ± 61..02	1464.46 ± 45.72
	F	1913.47 ±46.27	2134.17 ± 72.16	1413.21 ± 59.62	1535.04± 49.28	1346.65 ± 29.80	1188.55 ± 28.46
40 Wk BW (g)	M	3373.27 ± 46.23	3986.43 ± 62.17	3034.22 ± 52.84	2686.83± 79.20	2462.09 ± 28..05	2030.80± 34.83
	F	2432.46 ± 56.73	2896.38 ± 42.63	2192.67 ± 62.49	1604.11± 58.15	2021.17 ± 28.39	1701.00 ± 21.61
Age at first egg (days)		181.23	190.62	193.62	187.10	198.04 ± 1.70	201.26 ± 1.87
Egg production upto 40-weeks of age.		29.64	23.74	34.62	8.23	26.31	25.26

Breeding strategy for the development of location specific variety for North East region:

For evolving location specific new designer backyard bird, it is utmost essential to define a breeding strategy, which will be suitable for particular region. So first of all, we should know our assumption about the new designer backyard poultry bird, which can thrive well in low input technology (LIT) system.

1. The bird should be medium in size.
2. Having foraging habits and can thrive well with little or no supplementary feeding.
3. Average egg production should be more than 150 eggs per year.
4. Coloured bird with flight ability to save from predators.
5. Ability to adjust against harsh climate and resistance to most of the poultry diseases.

The indigenous germplasm have specific desirable traits like coloured feather, adaptability to local environment, disease resistance, foraging habits, broodiness, good carcass quality and able to produce under low input technology (LIT) system but poor in growth and egg production. However, the improved stock has desirable traits viz. higher body weight with good growth rate and egg production. So for evolving designer backyard poultry bird it is essential to merge the desirable traits of both the germplasm by crossing of improved stock with the native germplasm. The reciprocal progenies produced by this cross after evaluation, may be again crossed with improved germplasm to increase inheritance of improved germplasm to produce commercial crossbreds which would be more resemble to the native breed having coloured feather, better adaptability to local environment, moderate body weights as well as higher egg production and require low inputs. Using improved egg/broiler type breed/line depends on the purpose for which we want to develop backyard bird viz. egg, dual or meat purpose.

Improved egg/broiler type breed/line × Local native
(Higher body weight & high egg production) (Adaptability, disease resistance, foraging habit & good carcass quality traits)

↓

F_1 population ⟶ P1 P2 (Both direct and reciprocal)

Evaluation of both P1 and P2 populations (of F_1) under farm and field conditions and select the best population either P1 or P2 to be used for crossing purpose subsequently.

↓

Either P1 or P2 of (of F_1) × Improved egg/broiler type breed/line

↓

Crossbreds (Commercial backyard bird)
(Moderate body weight and egg production,
adaptability, disease resistance
and good carcass quality)
(Evaluation under farm and field conditions simultaneously)

The development of designer backyard bird, which would be suitable for rearing in low input technology (LIT) system, need the maintenance of separate parent lines for the native and improved germplasm. The Research Institutes should maintain the parent lines for both the germplasm and continuously improve both the germplasm separately generation to generation for the development of commercial variety by crossing the lines. After crossing, the fertile eggs have to be hatched in the Institute hatchery or may be given to private hatcheries. The chicks brooded and maintained upto 4-6 weeks of age by the key rearer. During this period, the chicks should be immunized for the deadly viral diseases of the poultry. After brooding and vaccination, the chicks will be distributed to the farmers for rearing in free range system.

Litter Management:

The shelter of poultry required locally available litter material, viz. Rice husk, paddy / wheat straw (2-3" size), saw dust, wood savings, dry leaves, groundnut kernels, sugarcane waste (broken into small pieces), dry grass or maize stalks. There may be problem of dampen in the litter in long rainy season of N.E region. This may initiate infection in the flock so mix hydrated lime @ 1.0 kg / 15-18 sq. ft. or Super phosphate @ 1.0 kg./ 15 sq. ft floor space. More lime may interfere the process of decomposition so restrict the quantity at optimum level. Super phosphate assist in preserving nitrogen level and no ammonia formation occurs. The litter is most important component, which after one year can convert into very good compost, which will be very useful to fertilize crops and vegetables. One poultry shelter consisting of 10 birds can produce approx. 300 kg poultry manure in a year.

The advantages of poultry litter: -

(1) Litter maintains its own temperature in all weather so that it helps to maintain the bird temperature uniformly and acts as insulator.

(2) Birds and eggs are safe with enclosed condition.

(3) Part of the feed requirement of poultry comes from the litter. The well built up litter supplies animal protein factor (Vitamins B_{12}) as in fish meal, meat meal and high in riboflavin (Vitamin B_2) content so there is no need to supplement feed for these factors. It can be mixed at high proportion into cattle feed (McArdle and Panda, 1970).

(4) Saving of labour because there is no need to clean floor daily.

(5) Some diseases are controlled by its action. Dry litter-reducing coccidiosis than the base floor sheds. Birds control body lice by dust bathing in the litter.

(6) It produce rich manure, has the value 3 times the cattle manure viz. 3% Nitrogen, 2% Phosphorus, 2% Potash, 0.68% Magnesium, 0.68% Sodium and 2.70% calcium plus trace elements. Nitrogen to organic matter ratio is approx. 1 to 33.

- 8-12 kg. Grains can be increased by the use of only one kg of nitrogen; an extra 6-8 kg grains for each kg of phosphorus and an additional 4 kg of grain for each kg of potash.

❖ 40 birds supply one ton of deep litter in a year, which is enough fertilizer to meet the full fertilizer needs of one acre of paddy or maize or 2 acres of sorghum or about ½ acre intensively cultivated crops such as vegetables. An extra ton of paddy per acre is possible from one ton of deep litter from 40 birds in a year.

Economics of Rural Poultry Production:

(A) Input: (For rearing 10 Gramapriya females).

(i)	Cost of 22 day old chicks (@ Rs. 10/- per chick)	Rs. 220.00
(ii)	Cost of brooding, vaccine and medicine (@ Rs.10/- per chick)	Rs. 220.00
(iii)	Cost of feed:	
	a. Upto 4 wks. of age full starter ration (@ 700 gm per chick so total 15.40 kg. and Rs. 20/- per kg.).	Rs. 308.00
	b. Supplement feed 25 gm. per bird upto 12 wks. of age for 20 birds and than 10 pullets upto 1.5 year i.e. 420 days, total 133 kg. @ Rs. 12/- per kg.).	Rs. 1596.00
(iv)	Cost of litter @ Rs. 30/- per bag (for 3 bags).	Rs. 90.00
(v)	Cost of housing material	Rs. 100.00
	Total	**Rs. 2,534.00**

(B) Output:

(i)	Sale of 10 cockerels at 12 wks. of age (Av. Live weight 1.2 kg. @ Rs. 120/- per kg.).	Rs. 1440.00
(ii)	Cost of eggs laid in a year (for 10 pullets @ 150 eggs in a year @ Rs. 4.00 per egg).	Rs. 6000.00

(iii)	Sale of 10 pullets after laying year (Av. Live wt. 2.0 kg. @ Rs. 100/- per kg.).	Rs.2000.00
(iv)	Cost of litter manure: (300 kg. manure with 3 % nitrogen, 2% phosphorus, 2% potassium and trace elements so 9.0 kg nitrogen @ 13/- per kg., 6.0 kg. phosphorus @ Rs. 36/- per kg. and 6.0 kg. potassium @ Rs. 14.16 per kg.).	Rs. 418.00
		Total Rs. 9858.00

Net profit = Rs. 9858.00 - Rs. 2534.00 = **Rs. 7,324.00**

(Rupees seven thousands three hundred twenty four) only.

Conclusion:

In the prevailing condition of North East region, the rural poultry production can be used as a tool for poverty alleviation in the rural areas. It can fulfill the nutritional requirements and help in the generation of rural employment. The type of bird and flock size of poultry may depend on the access of farmer's land, labour and capital. In this system it is easy to maintain birds without much labour and expenditure. The use of chemicals as well as pollution may be reduced so the system is environmental friendly and sustainable.

For overall development of the poultry in N.E. region, first; there should be location specific variety for the this region, which would be developed using local germplasm of the region. The newly develop birds would be having the advantage of adaptability and disease resistance along with high growth rate and egg production and thrive well in the region in comparison to other stocks procured from different regions of the country for rearing. Second; there is a need to develop cheaper balance ration using locally available ingredients so the farmers can afford to buy the feed for feeding as a minimum supplement to the birds. Third; supply of chicks on regular basis to the farmers of this region is most difficult task. Therefore, the Animal Husbandry Department of the concerned state Govt. should estimate the requirements according to actual need of the farmers in the state and should make the arrangement for supply of chicks. Fourth; training to the farmers is also required for the scientific management of the poultry production. It can help in increasing egg production and reducing the poultry diseases. If the state Governments of North East region emphasis on all these four points in their development plan, not only they can save good resource, which is going to other states for the purchase for eggs and chicks but would also help in livelihood improvement and poverty alleviation in the rural areas of this region.

References:

Annual Reports (2001-09). AICRP on Poultry Breeding (Rural Poultry Production). ICAR RC for NEH Region, Tripura Centre, Lembucherra – 799 210, Tripura West.

17th Livestock census. (2003) Department of Animal Husbandry and Dairying. Ministry of Agriculture. Govt. of India, New Delhi.

Malik, S. and Singh, N.P. (2009) Strategy for the sustainable rural poultry production in N.E.H. region. In: Food security and Sustainable Agriculture. PP: 278-288. Published by U-Day Publishers and Advertisers, New Delhi.

McArdle, A. Allan and Panda, J. N. (1970) A Poultry Guide for the villager. Department of Agriculture, Ministry of Food and Agriculture, Government of India, New Delhi.

Statistics of State / UT Animal Husbandry Departments (2003-04).

Viroji Rao, S.T. (2006) Research priorities and policy options for backyard poultry development to 2020. In: Proceedings of national Seminar on "Poultry Research Priorities to 2020" Nov., 2-3, 2006, PP: 244-249. CARI, Izatnagar-243 122.

* * *

5

INFERTILITY AND STERILITY IN DAIRY ANIMALS

D. P. Singh and D. C. Rai
Department of Animal Husbandry and Dairying
Institute of Agricultural Sciences
Banaras Hindu University, Varanasi (U.P.)

Introduction:

India has the largest Livestock population in the world after Brazil. The cattle population in 2003 was 7424,000 and 7976,000 in 2007. Livestock density of our country is 2.6 /hectare while it is 0.8/ hectare in the developed nation. Two third of Livestock owners are generally poor, small and marginal farmers and landless laborer owning only 30% of agricultural land. Ownership more evenly distributed amongst poorer section of society which is essential to achieve social justice.

The livestock industry of India contributes to 5.30 % of the GDP of India (2006-2007), whereas livestock contribute 30% of total agricultural production in India. Indian agricultural industry makes to the GDP of India is 17.1%. As per experts, "The contribution of Indian livestock industry was Rs.51Billion to the total export earnings of India in FY05, out of which the contribution of leather and meat products was 50% & 25% respectively". Livestock mainly stay in arid, semi arid, drought condition. Dairying in Meghalaya is of subsistence nature unlike in other regions of the country. The people rear cattle and buffalo mainly for meat and milk but in some pockets these species are preferred for draught purpose. Goat and sheep are kept by the poor section mainly for meat purpose. The major part of the milk produce is obtained from cattle followed by buffaloes.

In our country, about 75 million families are engaged in milk production. Milk production is predominately the domain of landless laborers, small and marginal farmers in India, who generally keep 1-2 animals under mixed farming system. These small holders have about 70 percent of the milch animals. Due to fragmentation of land the average land holding is declining, while the member of operational holdings is increasing. The average land holding is about 1.41 hectare per family which is expected to decline further. Out of 98 million holdings, about 30 percent rural holding are landless. The economic conditions of these landless farmers are mainly supported by the dairy animals which provide strength to the farmers for surviving, on the other hand the farmers which generally restrict themselves up to crop production. They experience a heavy loss due to which they go for the suicide as they unable to survive under scarcity. The livestock rearing for dairying is an additional occupation and a supplementary relationship with other farm practices of largely agrarian populace. The per capita availability of land is diminishing due to population escalation and excessive destruction of land for practicing traditional jhum cultivation. Livestock ownership is more evenly distributed with landless laborers, small and marginal farmers and approximately 75 percent of the total farm holdings fall in the category of small and marginal (average holdings of less than 1 hectare), hence, no favorable conditions exists for agriculture farming (Momin et al. 2009).

Cattle are an integral part of livestock agriculture for over 5000 years, producing draft power, milk, meat and hide. Around 2 Lakhs milch cows and buffaloes are spread over 5000 villages in Meghalya. About 1.64 Lakhs households are engaged in rearing small animals and ruminants and other allied activities. Most of the livestock rearers are below the poverty line. The returns from livestock sectors especially dairying and mixed farming is small and medium holdings are larger and highly sustainable (Momin et al. 2009). In dairy animals milk is the only thing which decides the profit and it is directly influenced by the fertility of the animals. Therefore, to get more profit from the dairy cattle's, the herd should have high fertility. On the other hand infertility is one of the most important economic losses in high-producing dairy herds. Losses are caused by retained placenta, metritis, anestrus, silent estrus, cystic follicles, repeat breeding, and abortions. Any of these conditions may affect only an individual or a large part of the herd. In some situations, modern feeding and management practices in large herds have predisposed to and accentuated infertility problems. The approach to solving the herd infertility problem should be as systematic and objective as possible. Even with these precautions, it is not always possible to determine the cause of infertility.

The economics of dairy is not based on milk production but largely on the function of reproduction efficiency. To explore the full production potential of dairy cattle, it is essential that they should posses a high reproductive efficiency. Reproduction disorders produce infertility or sterility in livestock. Infertility is a degree of reduced fertility which results in failure to produce or delay in producing live calf.

Cattle are deemed infertile when they are neither normally fertile nor completely sterile. Interest in bovine infertility increased with the introduction of artificial insemination in the 1950s and as the factors involved became known to farmers, herdsmen, physiologists and other workers (Roberts, 1956). The causes of infertility are many and can be complex (Arthur, 1982). They relate to Graafian follicle development and maturation, oestrus onset, successful coitus, ovulation, fertilization, implantation, and the development and delivery of the foetus and its membranes. Anything interfering with these routines, such as diseases, poor nutrition, inadequate herd management, hereditary and congenital factors, hormonal disturbances or environmental changes, makes the animal infertile, if only temporarily (Osmanu, 1979).

Ten to thirty percent of lactations may be affected by infertility and reproductive disorders (Erb and Martin, 1980), and 3-6% of the herd is culled annually in developed countries for these reasons. The extent of the problem is likely to be similar in the tropics, although extensive data are not available. In Zambia, for example, the National Council for Scientific Research (NCSR, 1970) noted that infertility is one of the major problems confronting the cattle industry, but that the extent and causes were still obscure. As reproduction plays a vital role in the propagation of the species. It assumes prime importance in livestock economics since reproduction is an essential prerequisite to livestock productivity. Global attention is therefore focused on developing newer techniques for enhancing reproductive efficiency and maximizing fertility in livestock breeding. Disorders of reproduction are great impediment in achieving the desired goal. It is therefore, imperative to pay equal attention to this problem and evolve suitable remedial measures to overcome the complication of reproductive disorders of livestock.

The term fertility denotes the ability to produce a viable offspring. Reproduction disorders results in a degree of reduction in fertility with delay in the normal frequency of annual calving rate, leading to infertility. Complete or total failure of reproduction ability in a female or male animal is termed sterility. Exact information on the impact of infertility on economic loss in Indian livestock is lacking. However, it is estimated that around 18-40 percent cattle are culled due to infertility or sterility, annually. Reproductive disorders (infertility) occurs in both sexes the female (cow) and male (bull). The term gynecological disorders are applied to female reproductive disorders. The term andrological disorders refer to disorders of male reproduction. Congenital causes of infertility are often inherited. They include developmental abnormalities of the ovaries, oviducts, uterus, cervix, vagina and vulva. Some are lethal; a few have a morphological significance and others a functional significance. Common morphological conditions include ovarian (gonadal) hypoplasia and aplasia, anomalies of the tubular genitalia, hermaphroditism, freemartinism, arrested development of the Mullerian ducts (White heifer disease) and double cervix (Lagerlof, 1963). Many of these abnormalities have been documented in zebu cattle by Perkins *et al.*(1954), Kodagali (1974), Chenna (1980) and Kumi-Diaka *et al* (1981). However, they are of little significance if an appropriate culling programme is practiced. Apart from other scientists Singh *et al* (2009) reported that the herbal drugs show significant effects on the reproductive status of the cross-bred heifers and it was observed maximum in treatment group during spring season, as shown in the following table.

Table: Oestrus response and oestrus intensity of the cross-bred heifers in different seasons

Seasons and treatment	**Oestrus Response**	**Oestrus Intensity**	**Percentage of Oestrus**
Spring Season T_1 (Control)	2 out of 6 cross-bred eifers	Both in intense heat	**33.33**
T_2 (Treated)	6 out of 6 cross-bred eifers	5 intense and 1 intermediate	**100**
T_3 (Treated)	4 out of 6 cross-bred eifers	2 intense and 2 intermediate	66.67
Rainy Season T_1 (Control)	0 out of 6 cross-bred eifers	------------- ------	**0**
T_2 (Treated)	3 out of 6 cross-bred eifers	2 intense and 1 intermediate	**50**
T_3 (Treated)	2 out of 6 cross-bred eifers	Both intense heat	33.33
Winter Season T_1 (Control)	0 out of 6 cross-bred eifers	------------ ----	**0**
T_2 (Treated)	4 out of 6 cross-bred eifers	2 intense and 2 intermediate	**66.67**
T_3 (Treated)	3 out of 6 cross-bred eifers	Both intense heat	**50**

Recently Haryana state Government have decided to start an innovative scheme titled "Reproductive Health Management to Optimize Fertility" to make the livestock infertility free. In this multipurpose animal camps would be organized in the state. As many as 1000 camps would be organized in the state during the current financial year (2010) through about 300 teams. Each team assisted by paramedical staff. Each Saturday would be fixed as camp day. Each day would organize about 30 to 40 camps in its jurisdiction;

the services which would be provided included general health examination and treatment, induction and synchronization of oestrus through different hormonal protocols.

Infertility and sterility

Infertility is the condition in which the animal is unable to reproduce successfully. While infertility refers to a temporary condition but when becomes permanent it will be called sterility. Infertility can be of different intensities and may last for varying periods. Corrective measures can overcome infertility among cows.

Range of Culling

In general low percentage of sterility occurs in younger animal (approximately 2 to 3 percent) but increases in case of older cows (approximately 5 to 6 percent), which need culling from the herd to percent loss of milk production, decrease in returns, wastages in feed thus resulting in improvement of herd.Jain (1979) reported that the sterility and reproductive troubles in domestic animals are causing incalculable economic losses in term of production and work. Anoestrum, repeat breeding embryonal or foetal mortality during pregnancy and calf mortality are the major problems of reproduction in dairy animals. In a survey on 20,000 adult animals, conducted at I.V.R.I. the incidence of infertility was found to be as follows.

Incidence of infertility in animals	Percent
Heifer	22.24
Cows	11.35
Buffalo heifers	10.28
Buffalo	6.25
Bulls	11.65
Buffalo bulls	03.65

The incidence of infertility and reproductive disorders may be even more in several populations. Sterility in some areas of Haryana indicates the following percentage of reproductive status in cows and buffaloes.

Reproductive status in cows and buffaloes	Percent
Anoestrus	52.40
Cystic ovaries	1.01
Pyometra	0.91
Metritis	8.90
Repeat breeding	12.00
Miscellaneous cases	1.09
Animal pregnant	21.17

A well managed dairy herd should have 65-70 percent of cows/ buffaloes conceived on first service with on average of 1.3 to 1.7 services per conception. There should be less than 10 percent of the cows with reproductive problems. The calving interval from one calving to the next should be between 12 to 14 months.

Asdell (1951) made a study of reasons of removal of about 2792188 cows in US from the Dairy Herd Association herds and summarized all the reasons of cow removal in descending order as follow:

Reasons of removal of cows from herds	Percent
1. Low production	33.49
2. Dairy purposes	23.39
3. Udder trouble	11.47
4. Sterility	08.26
5. Abortions	07.34
6. Death	05.04
7. Old age	02.74
8. Other causes	08.26

Sterility and abortions accounted to nearly 1/6 of the total causes of cow's removal. The total culled cows in U.S. out of 2792188 cows on the test (1951) constituted 21.8 percent.

For profitable dairying, all animals which are uneconomical or non-remunerative, maturing late, have longer service period and low milk production should be culled. Aged animals and calves from parentage of low productivity, longer calving interval and slow growth must be removed. Male calves when not required should also be culled and disposed off.

Index of reduced infertility

Most farmers know when the fertility level of their seeds falls below normal but it is not easy for the veterinary surgeon to make a defective statement as to when infertility exists. The following guidelines have been expressed by Walsley (1971) as being indicative of reduced fertility.

1. Calving interval (index) exceeds 400 days.
2. Calving conception interval exceeds 120 days.
3. Calving rate to first service is less than 50 percent.
4. Average number of service per conception exceeds two.
5. At least one-third of the females require more than three services per calf.

Factors effecting reproductive efficiency can be due to:

1. Inherited
2. Anatomical
3. Physiological
4. Psychological
5. Nutritional
6. Pathological, and
7. Managemental causes

Theses causes may result in either (I) failure of ovulation (ii) failure of viable sperm production (iii) failure of sperm transport and deposition at the proper time (iv) failure of spermatozoa and ovum to meet (v) failure of fertilization though gametes meet (vi) failure of implantation and (vii) abortion.

Types of infertility:

I. Temporary: Following main reasons are responsible for causing Infertility problems in the livestocks:

A. Accidental

1. Inflammation of genitalia: This condition develops in animals due to infection after laceration, bruising etc. in genital organ of female and male animals.

(a) Bruising: Bruises often induce pain, but small bruises are not normally dangerous alone. Sometimes bruises can be serious, leading to other more life-threatening forms of hematoma; a bruise is always caused by internal bleeding into the extracellular space, usually initiated by blunt trauma. It usually occur when the cattle genetalia gets damaged, the underlying soft tissue turns red or black and blue and the resulting discoloration in the genital tissues is called a bruise. Bruises often change in color largely dependant on the depth of the bruise, thus the color of the bruise may change from yellow to green. This change in coloration occurs because of chemical decomposition in the affected area of the genetalia. The change is thus a result of the removal of dead cells and the replacement of the tissues by the body itself. The likelihood and severity of bruising depends on many factors, including type, severity of the bruise, type of blow or impact that caused the injury and healthiness of affected tissues.

(b) Laceration: Laceration, an irregular tear like wounds caused by some blunt trauma, is a tear in the mucosal layer of genetalia which results from an injury. Most minor lacerations have minimal bleeding, minimal pain and no numbness or tingling at the site.

2. Perforation of the uterine and vaginal wall: A variety of complications of this sort may arise due to abnormal parturition or service by a large and vigorous male. Under abnormal condition the penis may be caught by the urethral orifice or enlarged Gartners ducts and the fistulae may be formed.

3. Prolapse of uterus and vagina: This condition is generally faced during advanced pregnancy or when approaching parturition. Generally the vagina and at times also the uterus gets reversed and protrudes out through the vulva and make parturition difficult. This condition is also known as Ballooned Vagina.

B. Nutritional

Cattle fed normally do not exhibit any lowered fertility due to nutritional causes because the nutrients need for reproduction are the same as those required for maintenance, growth and milk production. Fertility is affected only when there are serious deficiencies.

1. Carbohydrate: Experiment with low energy levels, at Pennsylvania (USA) have indicated delay in spermatogenesis, reduction in the number of spermatozoa produced and poor viability of the sperms. Similarly in females the oogenesis may decrease and may cause infertility in females.

2. Fat: As far as the fats in the diet are concerned, no planned experiments have been conducted on large animals and ruminants. With laboratory animals it has been indicated that the unsaturated fatty acids are essential for well being of reproduction. Their deficiency may lead to a variety of troubles, namely, lack of libido in males, reabsorption of the foetus, abortion or dead birth, etc.

3. Protein: Very little work has been done on the effect of protein deficiencies on reproduction in cattle. It is difficult to imagine a practical condition of management in which protein alone could be a limiting factor. It has been suggested that spermatogenesis is more satisfactory if the animal proteins are included in the feed but critical experimentation has not supported this.

4. Minerals: Of the minerals that limit the reproduction, phosphorus seems to be the most important. The amount of phosphorus needed by an adult cow for efficient reproduction is about 10 to 12 gm per day.

i. Phosphorus: Interferes with ovarian function, causing, in milder cases, failure of heat but not ovulation.

ii. There is no evidence that calcium deficiency can be responsible for reproductive troubles in cattles but the C/P ratio is important.

iii. Calcium/Phosphorus ratio: Limit 1:2 and 2:1 for maximum efficiency.

iv. Iodine: Iodine deficiency has been blamed for some reproductive trouble namely lack of libido and poor quality semen. This is probably related to the lowered metabolism due to failure of the thyroid gland to secrete enough iodine containing hormones, the thyroxin. Its deficiency may also lead to premature, weak or dead birth.

II. Permanent

A. Genetical

1. White heifer disease: This is condition of persistent hymen. It is said to be inherited and more common in exotic breed of cattle.

2. Failure of erection of penis: An ability to copulate due to unstraightened sigmoid flexure of the penis of the bull. This due to failure of the retractor muscle and is a heritable feature.

3. Impotentia Coeundi (male): Failure of the retractor penis muscles to relax and to allow the penis to extend from the sheath and inheritable.

B. Anatomical

Many anatomical abnormalities contribute to infertility. Some of them are genetic in origin. In the females, anatomical conditions like incomplete genital tracts, double cervix (fig.1.), adhesion of uterus, ovarian adhesions, absence of cervix, persistent hymens etc. are observed. Apart from this some abnormalities are caused by injury or previous infection. In the bull, anatomical abnormalities like absence or under development of testicals, cryptorchidism, absence of portions of reproductive tract, deformations of reproductive tract or even limbs and feets etc. results in the reduction in reproductive efficiency.

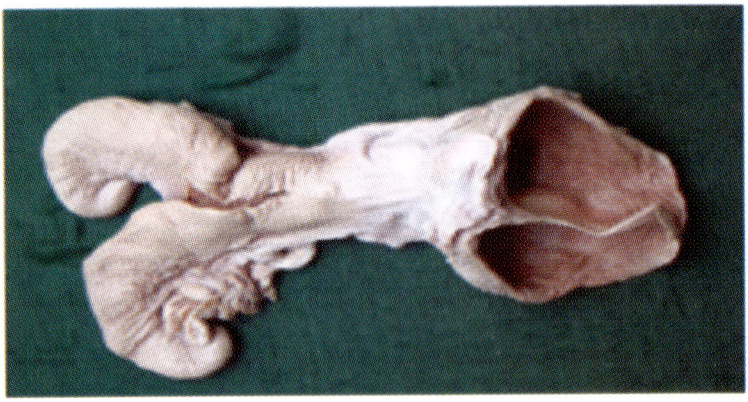

Fig.1: Double cervix

1. Cryptorchidism: It is the absence of one or both testes from the scrotum. This usually represents failure of the testes to move, or "descend," during fetal development from an abdominal position, through the inguinal canal, into the ipsilateral scrotum. Testes absent from the normal scrotal position can be:

- **a)** Found anywhere along the "path of descent" from high in the posterior (retroperitoneal) abdomen, just below the kidney, to the inguinal ring.
- **b)** Found in the inguinal canal.
- **c)** *Ectopic*, that is, found to have "wandered" from that path, usually outside the inguinal canal and sometimes even under the skin of the thigh, the perineum, the opposite scrotum, and femoral canal.
- **d)** Found to be undeveloped (*hypoplastic*) or severely abnormal (*dysgenetic*).
- **e)** Found to have vanished.

2. Scrotal Hornea: The decent of viscera into scrotum through the inguinal canal leads to the testicular atrophy by hampering their blood supply.

3. Impotentia Coeundi: Failure of the retractor peni muscles to relax and to allow the penis to extend from the sheath (Fig. 2).

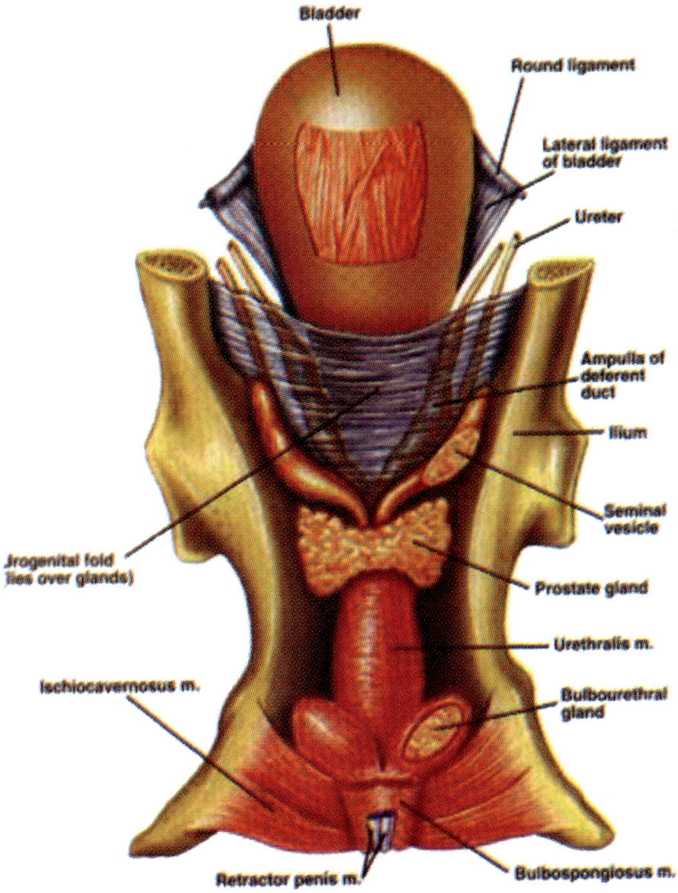

Fig.2: Penis coeundi

4. Free Martin: A freemartin or free-martin (sometimes martin heifer) is an infertile female mammal which has masculinized behavior and non-functioning ovaries. Genetically and externally the animal is female, but it is sterilized in the womb by hormones from a male twin, becoming an infertile partial intersex. Free martinism is the normal outcome of mixed-sex twins in all cattle species that have been studied, and it also occurs occasionally in other mammals including sheep, goats and pigs (Fig.3).

Fig.3: Freemartinism

5. Persistent Hymen (White Heifer Disease): White heifer disease associated commonly with cattle with all-white coats but occurs also in cattle of other colors and sporadically in other species. The heifer comes on heat and may breed but with difficulty and with straining afterwards. On rectal examination there is a distended uterus. The membrane may be reputable and the animal conceives but there may be other tubular defects and the animals are best discarded.

6. Incomplete Canalization and fusion of Reproductive duct: Sometimes the reproductive ducts like vas deferens, fallopian tube, uterine horns etc., are not properly canalized and may remain as solid structures. At times Mullerian ducts do not fuse or they may be the double vagina, double cervices, etc. These conditions may act as physical barriers.

7. Absence of parts: This feature is quite common amongst both the sexes and may be absent either unilateral or bilateral. The gonads or any part of the tubal genitalia may be absent. This excludes the animals from breeding either completely or partially. They are following types:

i. Aplasia of ovary
ii. Aplasia of oviduct
iii. Hypoplasia of ovary
iv. Hypoplasia of oviduct

C. Physiological

Most of the physiological causes of lowered fertility appear to be hormonal disturbances. Some of the major conditions are described as below:

1. Impaired sexual maturity: It may be due to pituitary dysfunction with regard to the gonadotrophic hormones (GTH). This may lead to underdeveloped or undeveloped condition of gonads and the accessory genitalia in both the sexes. Other hormones like thyroxin, somatotrophic, etc., which affect the general growth of the animal body may also affect sexual maturity.

2. Lack of gametogenesis and libido: Both these processes controlled by the FSH and LH component of the GTH, stimulate both spermatogenesis and oogenesis in the testes and ovaries respectively. The LH on the other hand stimulates the interstitial cells of the testes in male to produce the androgen which in turn induce the animal to exhibit its sex or sex drive otherwise called libido. The deficiency of any of these hormones may limit or retarded the reproduction in either sex.

3. Anoestrus condition: The anoestrus condition of the females may be due to a variety of causes of which the main ones are noted below:

i. **Infertile ovaries:** It may be due to genetic or nutritional causes. In these cases there will be no follicular development and the level of estrogen will be reduced so much that there hardly will be any oestrus.

ii. **Persistent corpus luetum:** The corpus luetum that remain functional beyond the normal period goes on producing progesterone which keeps the pituitary suppressed for the GTH activity. The absence of the FSH, therefore, does not stimulate the ovary to develop follicles which produce estrogen responsible for heat.

4. Silent or quiescent estrous: Frequently females go through the normal ovarian changes in the oestrus cycle but do not show the sign of heat and receptivity to males. Such cows and buffaloes are also known as silent or shy breeders. Most probably the deficiency of progesterone due to completely regressed corpus luteum may be the reason.

5. Short cycle and retarded implantation: This condition may be due to unhealthy or deficient C.L. This leads to the deficiency of the progesterone and frees the pituitary for GTH activity, thus stimulating the ovary for oogenesis and bringing the animal into heat even before due interval is over. The deficiency of the progesterone may also lead to improperly prepared conditions of the uterus for implantation of the foetus leading to no conception.

6. Cystic ovaries Nymphomania: In some cases single or multi-cysts may develop on the ovary and cause short oestrus cycles or prolonged oestrum. The animal usually will not conceive if bred at these abnormal heat periods. These animals are often referred to as chronic bullers and the condition is called nymphomania. Such cysts may be removed manually or treated with certain hormones, but the cow usually will not bred until she has passed one or two normal heat periods (Fig.4).

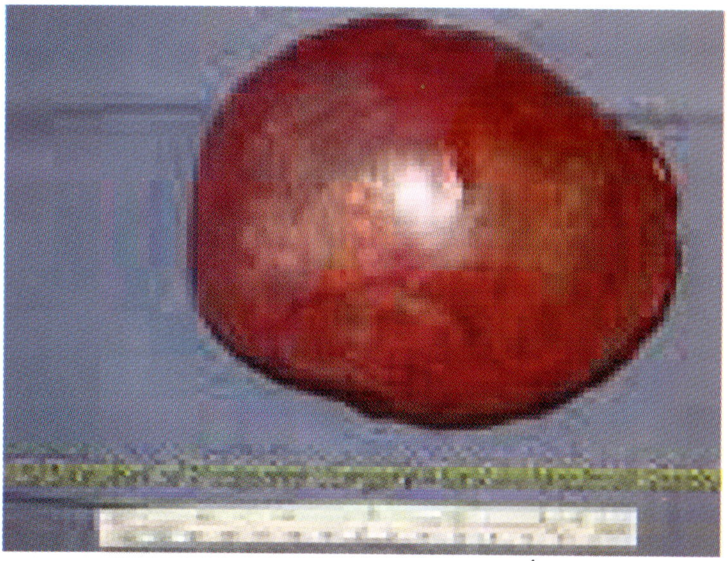

Fig.4: Cystic ovary

D. Psychological

The psychological disturbances lowering fertility rate have not been investigated much. However, general observation and limited experimental data point to fact that mistreatment and abuse may lower the fertility and conversely gentle handling and congenital surroundings enhance it. This is all the important when employing artificial insemination in which case the cow may become extremely nervous if not handled with patience and care. They nervousness on the part of cows and inexperience on the part of inseminators, both are found to cause lowered fertility. When the cow is extremely nervous, the adrenal medulla releases fright hormone epinephrine into the blood which counteracts the action of oxytocin. Oxytocin is released at the time of mating or artificial insemination which helps in transporting sperms from the cervix to the oviduct, where fertilization takes place, by causing strong contractions. When it is counteracted by epinephrine released due to fear and nervousness caused by rough handling and mistreatment, sperm transport is affected causing reduced fertility.

Mistreatment or unpleasurable experience on the part of the bull at the time of mating or semen collection may lead to psychological frigidity. Bulls may also develop a psychological inhibition against a particular surroundings, teaser or collection procedure in which case a change may overcome this problem. General mistreatment at times other than at the time of breeding has also been found to cause some lowering of fertility.

E. Pathological

Many pathological agents have been known to cause sterility and lowered fertility which can be extremely costly for the dairyman. Reproduction is affected only indirectly be general systematic diseases. But, some organisms cause a localized of the reproductive tract affecting fertility in a more direct way. These diseases may be transmitted by sexual contact or through other routes and may be caused by viruses, bacteria or protozoa. These come under specific pathogens.

On the other hand certain non-specific organisms which are not normally pathogenic may also cause conditions like bovine non-specific endometritis. Some mycotic infections lowering fertility have also been identified. These all comes under non- specific pathogens.

1. Specific

a) **Brucellosis (Bangs disease):** Brucellosis, also called Bang's disease, Gibralter fever, Malta fever, Maltese fever, Mediterranean fever, rock fever, or undulant fever is a highly contagious zoonosis caused by ingestion of unsterilized milk or meat from infected animals, or close contact with their secretions. Transmission from human to human, for example through sexual contact or from mother to child, is exceedingly rare, but possible. *Brucella* spp. is small, Gram-negative, non-motile, non-spore-forming rods, which function as facultative intracellular parasites that cause chronic disease, which usually persists for life. Symptoms include profuse sweating and joint and muscle pain.

b) **Bovine venereal trichomoniasis:** Trichomoniasis is a protozoal venereal disease among cattle that can cause abortions, low pregnancy rates and delayed or prolonged calving seasons.

c) **Vibrio foetus:** The causative organism, Vibrio fetus, is transmitted at the time of coitus. Most infected bulls will remain carriers indefinitely unless they are treated successfully. Some chronic carrier cows retain infection throughout pregnancy and after calving. One case in a herd can be enough to perpetuate the disease and infect clean bulls. When Vibrio fetus enters a herd, cows of all ages are susceptible. While 15 to 45 percent may conceive at first mating, other animals repeatedly return to service and conception may be delayed 3 to 6 months. In other herds, conception may be delayed 12 months or longer. A delayed oestrus cycle of 25 to 60 days, or perhaps more than 100 days, may follow the first service to an infected bull, but a delayed interval might follow any service. Later in the course of the disease, the fertility of cows which have already encountered the infection will be fairly satisfactory. Only the cows that have not been served previously by an infected bull and the heifers present a problem. Difficulties from each succeeding group of heifers may continue as long as an infected bull is used for service.

d) **Leptospirosis:** It is the disease caused due to *Leptospira Pomona or L. icterohaemogiae.* It is an acute febrile disease charterised by rapid lysis of red blood cells with resultant jaundice, haemoglobinuria, oedema and breathlessness. Lactating cows show sudden depression of yield, the udder secretion becoming pink or brownish. The mortality rate up to 33 percent occurred in calves but adult cattle show a much higher recovery rate.

2. Non-specific

i. **Vaginitis:** A severe disease characterized by vaginitis in cows and epididymitis in bulls occurs sporadically in eastern and southern Africa, where it is referred to as epivag. The disease is spread by natural mating. In the early stages of infection, cows have intense vaginitis characterized by reddened mucosae without ulcers, erosions, or vesicular lesions. A thick, creamy, white to yellow discharge develops. The infection spreads to the uterus and uterine tubes, and salpingitis and fimbrial adhesions frequently result in permanent infertility. Although epivag has been transmitted experimentally by transferring exudates, the cause is unknown. Inflammation of vagina. It is due to some organic infection (virus like organism). It may be a vesicular or a nodular venereal disease (Fig.5).

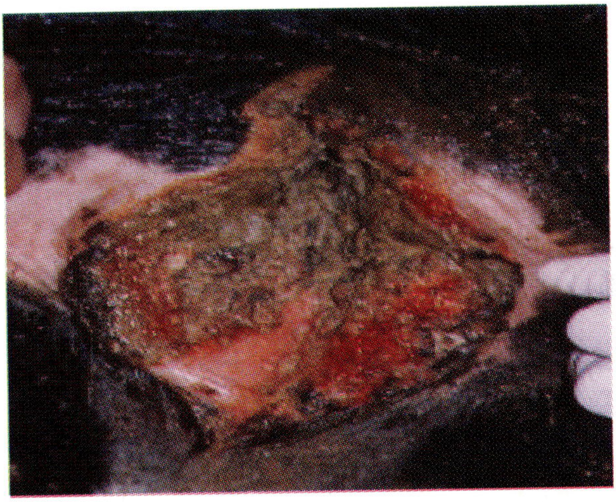

Fig.5: Vaginitis

ii. **Salpingitis:** Inflammation of fallopian tube. It occurs either due to viral or bacterial, fungal or protozoal infection present in female which leads to swelling may cause infertility. All such types infection occurs by mating with infected bulls and it may continue in the herd till infected bulls are being to mate with cows of the herd (Fig.6).

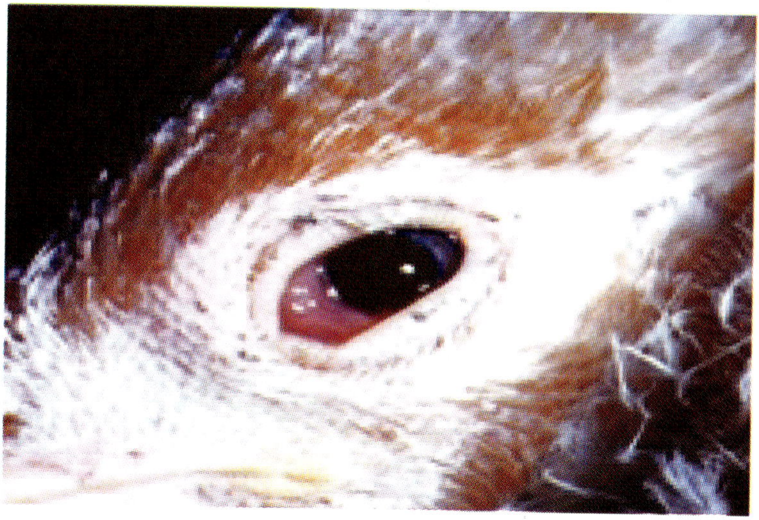

Fig.6: Salpingitis

iii. **Hydrosalpingitis:** Fallopian tubes become full of watery-fluid, and acts as spermotoxins.

iv. **Haematosalpingitis:** Fallopian tubes become full of blood. This kind of pathological condition occurs when fallopian tube is filled with water discharge.

v. **Pyosalpingitis:** Fallopian tubes become full of pus.

vi. **Metritis:** Metritis is the veterinary term for inflammation of the lining and the muscular wall of the cow's uterus and is generally caused by bacteria. It is painful and some cows will have raised temperatures and depressed appetites. In severe cases metritis will lead to shock, sudden death from toxicity or to peritonitis, because infection can transfer through the wall of the uterus in to the abdomen. In many cases metritis goes undiagnosed and untreated because "she's milking alright" but long term effects can be costly, often because of subsequent effects on fertility (Fig.7).

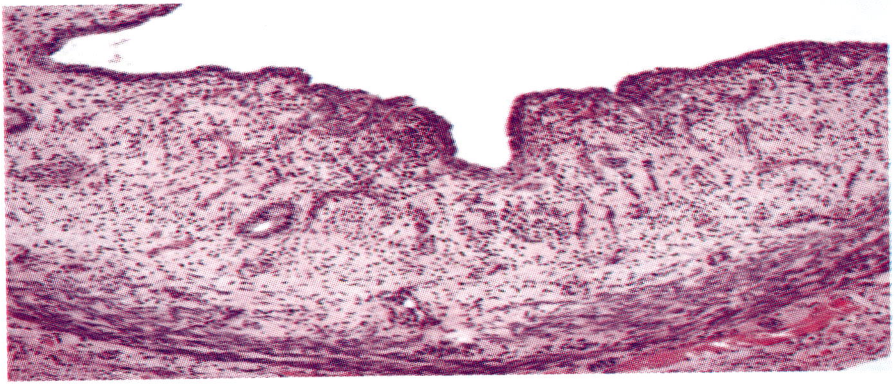

Fig.7: Metritis

vii. **Cervicitis:** Cervicitis, a common infection of the lower genital tract, is the inflammation of the cervix. Inflammation may be caused by infection from certain sexually transmitted diseases (STDs) or by injury to the cervix from a foreign object inserted in the vagina, from birth control devices such as the cervical cap or a diaphragm, or by cancer (Fig.8).

Cervicities symptoms include a red and inflamed cervix with an unusual discharge

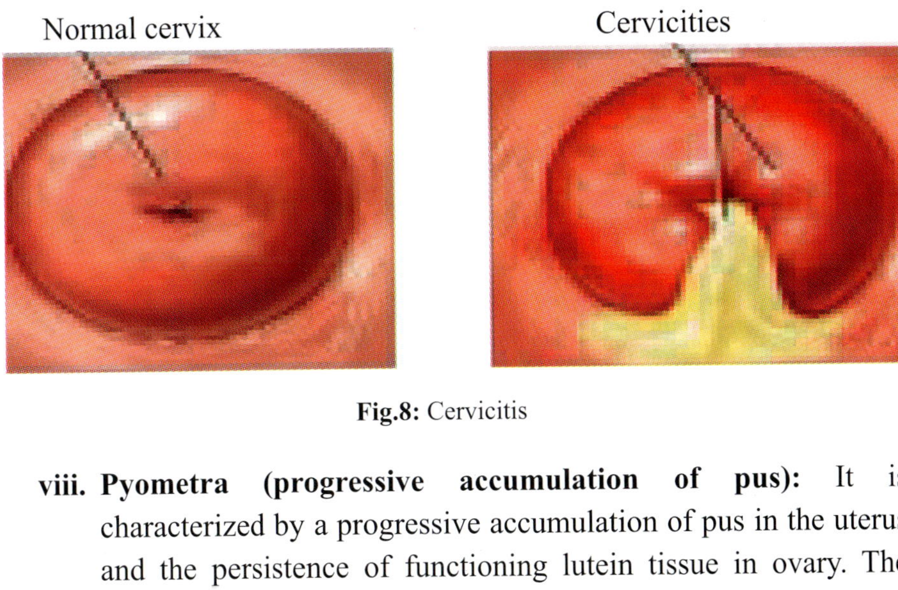

Fig.8: Cervicitis

viii. **Pyometra (progressive accumulation of pus):** It is characterized by a progressive accumulation of pus in the uterus and the persistence of functioning lutein tissue in ovary. The word "pyometra" is derived from Latin "pyo" meaning pus and "metra" meaning uterus. The pyometra is an abscessed, pus-filled infected uterus. Toxins and bacteria leak across the uterine walls and into the bloodstream causing life-threatening toxic effects, without treatment death is inevitable. It is characterized by the accumulation of purulent or mucopurulent exudate in the uterus. In cows, it is invariably accompanied by the persistence of an active corpus luteum and interruption of the estrous cycle (Fig.9).

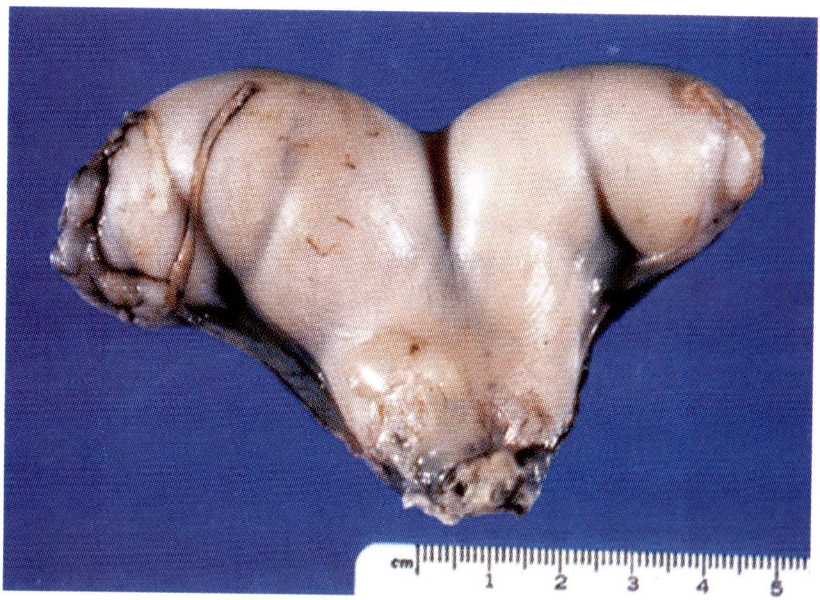

Fig.9: Pyometra

Remedies:

Most producers experience a herd problem with reproductive performance at some time. This may occur in the absence of any readily apparent change in management. When reproductive efficiency declines, the dairy producer should work with the herd veterinarian, artificial insemination (AI) technician, feed company representative, country extension agent and other resources people to troubleshoot the causes and determine solutions to the problem. The following material identifies the major types of herd infertility problems and their possible causes and gives suggestions for their prevention and control.

Besides this there is following points to be kept in mind by all dairymen for better production and may able to reduce infertility:

(i) Increased knowledge about the principle causes of reduced fertility is essential. Increases in milk yield have been at the expense of reduced fertility in dairy cows and although diet has a major impact, the precise interaction between nutrition and reproduction still needs to be characterized in dairy animals.

(ii) On the male side, bulls are selected for optimal freezability of semen; however, there is as yet no reliable predictor for semen fertility. Methods for accurately predetermining the sex of both semen and embryos will revolutionize the dairy.

(iii) If natural service is used, test for vibriosis and Trichomoniasis. If present, discontinue natural service and cull the bull. Use a vaccine for protection against vibriosis.

(iv) Use high fertility bulls.

(v) Carefully observe estrous onset and duration. Breed 12 hours after initial observation. On long-estrous cows (more than 24 hours), inseminate a second time 12 to 18 hours after the first insemination.

(vi) Use extreme care in handling the reproductive tract of possibly pregnant cows.

(vii) Eliminate cow access to swamps, ponds or standing water.

(viii) When an abortion occurs, submit samples (the entire fetus and at least a portion of the placenta including a cotyledon in separate plastic bags) to a reference laboratory for evaluation. Draw blood from an aborting cow at the time of the abortion and again three weeks later for serologic testing for brucellosis, leptospirosis, IBR and BVD.

Retained Placenta

When a freshing cow fails to expel her placenta (after birth) within 12 hours after calving, the condition is known as retained placenta. Incidence of retained placenta in dairy herds should not normally exceed 8 percent.

Possible Factors Involved:

1. Specific infectious such as **brucella, leptospira, campylobacter, infectious bovine rhinotracheitis (IBR)** and others can result in retained placenta. These infections may cause abortion but can also cause retained placenta following delivery at term.
2. Non- specific infections by a wide range of bacteria and virus that occur during pregnancy or at calving can be associated with retained placenta.
3. Twin births and abnormal deliveries, including prolonged or difficult deliveries or caesarian sections, are often followed by placental retention.
4. Deficiencies of selenium, vitamin A or vitamin E may cause higher than normal incidence of retained placenta.
5. Over conditioning of dry cows due to excess energy intake and/or prolonged dry period is often associated with retained placenta.

Troubleshooting and Control suggestions:

1. Test for specific infections use blood tests as well as bacterial or viral cultures in diagnosing specific infections. If an infection is identified, treat, vaccinate or cull infected cows as indicated.
2. Minimize exposure to non-specific organisms by keeping calving areas clean and well bedded. Calves on grass if practical, don't use maternity pens for any other purposes.
3. Breed heifers to bull with a record of calving ease (Fact sheet IRM-16). Observe freshing cows and heifers closely. Provide assistance if hard labour continues for over 30 minutes without

progress. If assistance is required, do so in a clean, gentle manner.

4. In selenium-deficient areas of the U.S., provide supplemental selenium as dry feed or injection to all dry cows. Take tests to determine selenium status. Provide cows with fresh forage as pasture or green chop for at least 4 to 5 weeks each year. Provide about 160,000 units of vitamin A (One milligram of carotene is equivalent to 400 units of vitamin A) from all sources (natural and supplemental).
5. Avoid over conditioning at drying off time have cows in about the body condition in which they should freshen. Limit access to high-energy feeds such as corn silage or grain during the dry period.

Metritis

Infection of the uterus is known as metritis. Cows normally have a red-to-brown discharge during the first two weeks after calving. If discharge persists beyond two weeks or if the discharge is foul-smelling, this is evidence of metritis.

Possible Factors Involved:

Many cows with retained placenta will develop metritis (see factors involved in retained placenta). Injury to the reproductive tract can occur at the time of breeding or uterine treatment. Contamination of the reproductive tract can happen at calving when cows and heifers are highly susceptible to infection. If the calving area is unclean or if assistance and/or treatment around calving time is unsanitary, metritis is a probable result. Use of uterine boluses can cause a sterile pus condition due to body reaction to materials contained in some boluses. Selenium or vitamin E deficiency may be associated with metritis. Over conditioning may predispose cows to many health problems at the time of calving or during early lactation; these include retained placenta, metritis, acetonemia and displaced abomasums.

Troubleshooting and Control suggestions:

1. If incidence of retained placenta is greater than 8 percent, see control measures listed earlier. Many of the same conditions that predispose cows to retained placenta contribute to development of metritis.

2. Calve cows in a clean place.

   a. Use reasonably clean paddocks or pastures in good weather. Locate calving area to permit frequent observation.

   b. Clean, sanitize and re-bed maternity pens after each calving. Use long term bedding instead of sawdust if possible.

   c. Rest calving sites for one to two months when metritis or calf infections are widespread. Calve cows in a new, clean site.

   d. Keep maternity pens free from all animals except calving cows.

   e. If assistance at calving time is required, use clean, disinfected instruments; tie tail to front leg; wash vulva and surrounding area with a mild soap; wash arms and hands before assisting at delivery, use a non-irritating lubricant; work with the contractions of the cow; and don't exert excessive pressure. Be sure that the position of the calf is normal before attempting to pull the calf. Seek veterinary assistance if possible of the calf or size of the calf relative to dam's size indicates a problem delivery.

3. Follow your veterinarian's recommendations on when to treat metritis and retained placenta and which medications to use.

4. In selenium-deficient areas, add supplemental selenium to milking and dry cow rations or give selenium injections.

5. Avoid over conditioning during late lactation and dry period, while maintaining adequate, balanced vitamin and minerals intake.

6. Have all cows examined by your veterinarian two to six weeks after calving to diagnose injury or infection.

7. Follow your veterinarian's recommendations for treatment of metritis. Avoid over treatment. Avoid routine medication of the uterus unless a cow is known to be infected.

Cystic Ovaries

Ovaries cysts are structures, usually greater than one inch in diameter, that persist on one or both ovaries for ten days or more. Fertility in cystic ovaries cows is reduced due to hormonal changes, changes in uterine tone and in many cases failure to release an ovum (egg).

Possible Factors Involved:

1. Excessive calcium intake or wide calcium phosphorus ratio. Total dietary intake of greater than two parts calcium to one part phosphorus may lead to increased incidence of cysts.
2. High estrogen intake whether given by injection, through fresh legume forage or from some mold toxins, may increase the incidence of cystic ovaries.
3. Genetic predisposition.
4. Stressful conditions or health problems at calving or early postpartum.

Troubleshooting and Control suggestions:

1. Have forages analyzed, including calcium and phosphorus analysis. Check the feed program to insure that the calcium-phosphorus ratio is between 1.5:1 and 2:1 in the total diet. Include all forages, grains and free-choice minerals in estimating minerals intake.
2. Avoid use of injectable estrogen products. These are potent drugs for use only by or under the direction of a veterinarian. Analyze feeds suspected of containing zearalen one or other mold toxins. Do not use feedstuffs containing high levels of other estrogenic

mold toxins or plant estrogens to breeding stock or limit feed to reduce estrogen intake.

3. Since genetic predisposition to ovarian cysts can occur, reduce herd incidence of cystic ovaries by selective culling of cows known to produce cystic daughters and avoid use of bulls known to sire cystic daughters.

Anestrous

Anestrous or failure to show signs of estrus (heat) is in most cases a failure to detect estrus. In some of the studies cows reported to be anestrous, 90 percent were determined to be cycling but not detected in estrus and only 10 percent were truly anestrous (no ovaries activity).

Possible Factors Involved:

1. Undetected estrous signs in cows with normal ovaries activity resulting from:
   a. Inadequate estrous detection since 66 percent of estrous signs are shown between 6 p.m. and 6 a.m. Cows with short estrous (less than 12 hours in length) may be missed even with twice a day estrous detection. This is particularly true when cows are observed for estrous during a time when they are unlikely to exhibit standing behavior.
   b. Inadequate animal identification and/or inadequate records (Fact sheet IRM-4).
   c. Lack of opportunity for cows to express estrous, i.e., cows not turned out, slippery footing, lameness or stiffness, group too small for adequate interaction.
2. True anestrous, lack of ovarian activity, caused by:
   a. Anemia-due to anaplasmosis, internal or external parasites, deficiency of protein, iron, Copper, cobalt or selenium, phosphorus deficiency.
   b. Energy deficiency, cows losing flesh due to high production and/or underfeeding.

c. Low hormone levels associated with prolonged feeding exclusively on stored feeds.

d. Cystic ovaries (70 percent of cystic cows are anestrous).

e. Pyometra or pus in the uterus.

3. Quiet estrous or silent estrous- normal ovarian activity with little or no signs of estrous. Most of the factors associated with true anestrous (anemia, phosphorus deficiency, energy deficiency and low endocrine levels) can also be associated with quiet estrous.

Troubleshooting and Control suggestions:

1. Maintain adequate reproductive records. Record all estrous dates, examination dates and findings, unusual events such as difficult calving or retained placenta and treatments.
2. Closely observe cows for estrous.
   a. Observe cows at least twice and preferably three times a day for estrous. Make such an estrous detection time is in the evening.
   b. Observe cows for at least 20 minutes at each detection time.
   c. Don't let any other farm chores distract from estrous detection. Observe cows when they are not occupied with eating, being milked or other activities.
   d. Provide a non-slippery surface for estrous detection, so cows are most likely to exhibit signs of estrous.
3. Check problem cow for anemia take unclotted blood samples from 12 to 21 early lactation cows. If anemia is present, check for internal and external parasites. Check blood for iron, copper and selenium status. Treat for anemia if detected. Correct predisposing causes.
4. Submit forage samples for standard and mineral test. Seeks assistance to develop a feed program that meets herd nutritional requirements.
5. Have cows examined for uterine infection, cystic ovaries and for evidence of ovarian activity. Have a veterinarian palpate all cows at least once between 15 and 45 days after calving.

6. Whenever possible, have cows in a weight gaining condition as desired breeding time approaches.
7. Control conditions around calving time and early lactation may contribute to anestrus problems (retained placenta, metritis, and ketosis). Provide access to fresh forage for at least 4 to 6 weeks each year.
8. Have pregnancy exam done 40 to 60 days after breeding.

Repeat Breeders:

A reasonable goal for first-service conception rate is 50 percent. After two services the cumulative conception rate should be around 75%. Cows requiring three or more services before conception or culling are generally designated as repeat breeders.

Possible Factors Involved:

1. Embryonic or early fetal mortality caused by:
   a. gross overfeeding;
   b. excessive manipulation of the reproductive tract by rectal examination;
   c. Trichomoniasis or vibriosis (Either of these venereal diseases can cause early embryonic death or repeat breeding. These diseases are spread by natural service);
   d. Leptospira and possibly virus infections of the reproductive tract;
   e. ureaplasm, mycoplasm, haemophilus and many other infectious agents which produce a low grade infection that either prevents conception or results in early embryonic death. Affected cow may or may not show abnormal discharge.
2. Breeding too early or too late in relation to time of ovulation.
3. Use of low fertility sires.
4. Use of semen damaged in storage or handling.
5. Poor insemination technique.
6. Serious imbalance or deficiencies of vitamins or minerals.

Troubleshooting and control Procedures:

1. If natural service is used, test for vibriosis and trichomoniasis; if present, discontinue natural service and cull the bull. Use a vaccine for protection against vibriosis.
2. Test for Ieptospira, IBR, BVD. Use available vaccines for protection against these organisms.
3. Avoid gross overfeeding of grains, especially after breeding.
4. Provide cows with access to fresh forage such as pasture or green chop for at least 4 to 6 weeks a year.
5. Carefully observe estrous onset and duration. Breed 12 hours after initial observation. On long-estrous cows (more than 24 hours), inseminate a second time 12 to 18 hours after insemination.
6. Have a veterinarian examine repeat breeders cows for presence of endometritis (low-grade uterine infections), delayed ovulation or other abnormalities.
7. Use high fertility bulls. Have semen handling and insemination technique checked by competent AI personnel.
8. Semen viability of representative straws can be analyzed.

Abortions:

Causes of abortion can be either infectious or non-infectious. In about 70% of abortions, the cause can not be determined even with careful laboratory examination. In the absence of specific infections, a heard abortion rate of around 3% is normal.

Possible Factors Involved:

1. **Genetic defects-**A severely abnormal embryo or fetus is likely to be absorbed or expelled in early pregnancy.
2. **Multiple fetuses-**Abortion rates are higher in multiple-fetus pregnancies.
3. **Injuries-** In early pregnancy (less than 60 days) excessive manipulation of the reproductive tract at the time of rectal examination can cause abortion. Passage of an insemination or

infusion instrument into the uterus of a pregnant cow can cause abortion. Severe injuries in late pregnancy such as falling on a hard surface can induce absorption.

4. **Specific infections-** Brucellosis, Leptospirosis, IBR, BVD, Lysteriosis, Vibriosis, Trichomoniasis, Mycotic infections and others can cause abortion.
5. **Toxicities-** Acute nitrate, cyanide, silo gas and some weeds can cause abortion.
6. Drug-induced abortions- Cortisone-type drugs and prostaglandins, among others, can cause abortion.

Troubleshooting and Control Procedures:

1. Avoid injuries. Groove slippery concrete surface and/ or apply nonslip materials to floor. Keep cows off icy areas.
2. Use extreme care in handling the reproductive tract of possibly pregnant cows. On repeat services do not allow the insemination rod to enter the body of uterus. Discuss with your veterinarian the minimum number of days since last service at which he or she feels capable of detecting pregnancy without risk to the fetus.
3. When an abortion occurs, submit samples for laboratory diagnosis. Place the entire fetus in a plastic bag. Put the placenta or at least a portion of the placenta including a cotyledon in a separate plastic bag. Submit to a reference laboratory for evaluation. Draw blood from an aborting cow at the time of the abortion and again three weeks later for serological testing for Brucellosis, Leptospirosis, IBR, BVD.
4. Eliminate cow access to swamp, pond or standing water.
5. Submit for laboratory analysis any suspected feed and water. Toxicities severe enough to cause abortion are generally short-term access to high levels of a poison or toxin.
6. Most abortion-inducing drugs, such as cortisones and prostaglandins, are prescription dugs.

Allow use only by a veterinarian or on his or her instruction.

When herd reproductive problems occur, they can have a major economic impact. Good records and a good working relationship with the herd veterinarian and other resource people minimize the chances of reproductive problems. If they do occurs, use records, resource people and a systematic troubleshooting approach to solve those infertility problems.

Precautions:

1. Supply of balanced ration to farm animals.
2. Proper recording of History sheets and Pedigree records of farms animals.
3. Regular testing of Dairy Herds for tuberculosis, brucellosis and vibriosis etc.
4. Regular vaccination of dairy herds especially for such diseases responsible to cause infection.

Conclusion:

Economics of Dairy based on the functions of reproductive efficiency. So, It is important that the animals should be free from infertility or sterility problems which can be achieved by reducing infertility to have maximum production.

Acknowledgement:

Authors thankfully valued the contribution of **Mr. Shailesh Bhushan Srivastava** *(Research Scholar)* and **Mr. Surendra Kumar Gupta** *(Research Scholar)* of the Department of A.H. & Dairying, B.H.U., Varanasi (U.P.).

Reference:

Arthur, G.H. (1982). Veterinary reproduction and obstetrics. 5th edition.bailleire Tindall, London, UK.616pp.

Cheena, R.M.C. (1980). A case of uncommon voluminous vaginal fibroma in a nondescript cow. Indian Veterinary Journal57:861.

Erb, H.N. and Martin, S.W. (1980). Interrelationship between production and reproduction diseases in Holstein cows. Journal of Dairy Science 63:1911-1917.

Jamester, W. Momin and B.K. Mishra (2009). A Review on Livestock and Dairy Development in Meghalya, Bioved, 20(1, 2): 127-137.

Kodgali, S.B. (1974). Report on study of infertility in cattle 1965-69. Veterinary bulletin 45:1285 (Animal Breeding Abstracts 43:3369).

Kumi-Dika,J.; Ogwu,D.and Osori,D.I.K.(1981).Significance of atrophic ovaries in livestock production in northern Nigeria. Veterinary Record 108:277-278.

Lagerlof, N. (1963).Hereditary factors in infertility in cattle in: Infertility in livestock Animal Health Branch Monograph No.5.FAO (Food and Agriculture Organization of the United Nations), Rome, Italy.pp.66-77.

NCSR (National Council for Scientific Research 1970). Factors affecting the calving rate in local breeds of cattle in Zambia. Animal Production Research Report NCSR/TR 7. NCSR, Luska, Zambia.20pp.

Osmanu, S.T. (1979). Studies on bovine infertility at the Agriculture Research Station (Legon) over half of a decade (1972-77). Ghana University, Department of Animal Science Studies, Legon, Ghana.82p.

Perkins, J.R.; Olds, D. and Seath, D.M. (1954).Study of 1000 bovine genetalia. Journal of Dairy Science 37:1158-1163.

Roberts, S.J. (1956). Veterinary obstetrics and genital diseases. 1st edition. Edwards Brothers, Ann Arbour, Michigan, USA.

S. K. Singh, D. P. Singh, D. C. Rai and Shailesh Bhushan (2009).Efficacy of compound herbal drugs on reproductive status of cross-bred heifers. Indian J. of Environment and Ecoplanning. 16(2-3):593-600.

* * *

6

ORGANIC DAIRY FARMING: SCOPE AND LIMITATIONS

Dr. Sunil Kumar
College of Forestry and Hill Agriculture, G.B.P.U.A.&T.,
Hill Campus, Ranichauri, Tehri Garhwal, Uttarakhand, 249199,
E-mail: sunilgbpuat@gmail.com

ABSTRACT

India has undoubtedly made tremendous progress in dairying, but mostly in numbers quantities. What now our country requires in a qualitative improvement in the entire food chain i.e, from primary production to consumer so that India can establish itself as a reliable supplier of dairy products confirming to international standards and quality assurance system. Today the consumer asking for value for money in terms of quality of products, besides consumer is becoming conscious for the safety and suitability of foods for consumption. An increasing number of people, even in poverty stricken developing countries are looking for quality added products, and they do not mind paying a little extra for quality assurance. It is in this context that along with organic farming, the question of organic dairy has come up before us for consideration. The low external inputs based Indian dairy sector has better opportunities to convert to organic dairy production, since majority of the farmers in our country are organic farmers not by choice but by default. To get organic milk, it is precisely declared not to use pesticides, insecticides or hormones besides organic farm practices.

Keywords: Organic dairy, farming, scope, constraints.

Introduction:

The organic farming movement is commonly agreed to have begun in the 1940s in England with the writings of Sir Albert Howard, who learned about organic practices in India during the 1920s. The organic movement is built on a fundamental principle: healthy soils lead to healthy crops, healthy animals, healthy humans, and a healthy planet. Organic crop and livestock production focuses on building soil organic matter and biology to create a sustainable, dynamic environment for producing healthy food and feed. Organic agriculture is also seen as a way to sustain and support family farms in preference to faceless, ever-expanding mega- and corporate farm models.

Whereas organic fruits, vegetables, grains, and some livestock have long been mainstays of the organic movement, organic dairy is a relative newcomer. Organic dairy surged into the organic marketplace in the 1990s, establishing itself as a major category. The success of organic dairy can largely be attributed to several critical events, including a response to Monsanto's introduction in 1994 of genetically modified or recombinant Bovine Growth Hormone (rBGH). The proliferation of rBGH use—coupled with increased consumer awareness of genetically modified corn, soybean, and other crops treated with an array of synthetic pesticides being fed to livestock; the feeding of slaughter by-products to ruminants and concerns about mad cow disease; and the increased use of synthetic medications including hormones, antibiotics, and steroids—have encouraged many consumers to seek organic dairy products. These consumers have come to rely on the assurances of certified organic dairy as a trusted source of unadulterated dairy products.

At the same time, the organization and marketing efforts of producers and manufacturers of organic foods have established an infrastructure and market presence that makes high-quality organic dairy products available, affordable, and desirable in both specialty natural- and mass-market groceries. Organic dairy products are often viewed as "gateway products", in that consumers will make their first forays into organic purchasing by buying organic dairy products, eventually increasing their allegiance to organic products as they become increasingly food savvy.

Defining Organic Dairy:

According US Department of Agriculture (USDA) organic milk is define as milk from dairy animals that have been extensively fed organic feed, have not been treated with synthetic hormones, are not given certain medication to treat sickness, and are held in pens with adequate space. According to organic consumer association organic milk is a milk produce without synthetic chemicals, hormones or antibiotics. International organization organic animal husbandry and organic farming is introduced as the new trend in the field of agriculture sector. Their product is safest in market. Consumer is inclining to organic product and ready to pay higher prices as safest products. Organic animal husbandry may be defined as the production of livestock under define standard. Greater attention is given to animal welfare without using growth promoters, animal offal or any other additives and at least 80% of feed must be grown according to organic standard without using artificial fertilizer or pesticide on the fodder crops. Many areas in India were crops residues, local grass, legumes and tree leaves are being use to grow animals. For treatment of animals farmers use locally available herbs and traditional Ayurvedic methods. These farmer practices animal Husbandry through organic inputs. Although, products produce under such system have so far not been define as organic products but by all means they are genuine organic products. Now it is right time to classify these practices accordingly. These will go a long way in alleviating poverty and raising living standard of poor villages.

Ministry of commerce and industry Government of India New Delhi released National programme for organic production (NPOP), on 8th May

2000 and India organic logo released on 26th July 2002, for certification of Indian organic products. Under these programme ministry of commerce is promoting organic farming for export purpose and has established a regulatory mechanism which fixes standard for organic cultivation, accreditation of certification agencies and inspection etc. these standard are group under 6th categories; conversion, crop production, animal husbandry, food processing and handling, leveling, storage and transport. The certification of organic animal husbandry and dairy production standard ensure the management of animal environment. It takes into accounts behavioral needs of animals and provides for:-

1. Sufficient availability of sunlight.
2. Fresh air and spaces for free movement.
3. Protection against harsh environmental conditions i.e. Sunlight, rain, winds etc.
4. Sufficient availability of resting and laying area and ample access to feed and fresh water and
5. At least 80% fodder and feeds should be grown on organically managed own farm. Organic animal husbandry without fodder farm cannot be allowed.

It has a vast reservoir of indigenous technical knowledge in animal husbandry and milk production that has potentials to be effectively used in organic milk production. Organic milk production is yet to make forward steps though organic agriculture has made significant steps in international export arena. Some potentials area can be identified like North East India where traditional methods can be easily modified as per organic standards. Indigenous animal breed and indigenous animal husbandry practices deserved attention in a sense that these are ideal for organic animal husbandry and milk production.

Contamination of milk:

Indian milk are reported to contain different biological and chemical contaminations which could be introduce in various stages of production to processing and packaging. The dairy animals are fed on chemical fertilizer pesticides and growth promoters treated fodder and agriculture byproduct and crop residue. The residue present in animal feed not only has unwanted effect on health of animals but the residue and metabolites get accumulated in tissue and also secreted through normal milk secretion.

Source of contaminants in milk:

A. Biological

i. **Microbial agent-** Milk is a good medium for multiplication and propagation variety of pathogens. Pathogen enter into processes milk and milk product from unhygienic dairy equipment processing unit, dairy premises water, packaging under free atmosphere, personnel handling and from farm level such as cow teat, feces, urine as well as moist bedding material.

ii. **Somatic cell in milk-** somatic cells are the normal cellular infiltration of milk from peripheral circulation during lactation period. These cells include epithelial cells, macrophages, neutrophils and lymphocytes. The level of somatic cells count (SCC) is an indicator of udder health and quality of milk. The developed countries are giving more emphasis on SCC of milk for quality milk production and its export.

iii. **Mycotoxin-** The residue of mycotoxin in milk and milk product occur due to consumption of contaminated feeds stuff by dairy animals. Fungi although beneficial but some of them produce highly poisonous metabolites called mycotoxins. The contamination of dairy products occurs from fungal growth and for fermentation or an intestinal fungal growth lead to production of mycotoxin.

iv. **Natural plant toxins-** Consumption of phytotoxicated milk may induce poisoning in human or in neonates.

B. Veterinary drugs

i. Antibiotic- The antimicrobial drugs administered by either also cause to severe reaction to consumers. Antimicrobial residue in milk also lead to partial or complete inhibition of acid production by starter culture, inadequate ripening/fermentation and aging of cheese and also cause defects of flavor and texture of the products.

ii. Hormone- The use of hormone as growth promoter is approved in some countries was the natural steroids hormone only in milk production purpose. However safety concern regarding hormone residue in meat and milk has been raised by other countries residue problem in milk could be avoided if these hormone are used judiciously. The use of this hormone is not permissible if potentials risks are known to the health of consumer and livestock involved.

iii. Disinfectant- disinfectant are use in various stages of milk production collection, processing and packaging. Residues of disinfection and the sanitizers may enter into milk if cleaning disinfection draining and rinsing procedures are improperly conducted. Acid and alkali detergent cause local irritation on the skin.

C. Chemicals

i. Pesticide - It is reported that Indian milk and milk products contain detectable levels of pesticide residue. Since the development of synthetic pesticide consumer are more exposed to pesticides that pose public health hazard. Milk and milk product are more vulnerable to pesticide residue owing to its high fat content. The widespread use of pesticide persistence in environment and biomagnifications potentials of pesticide has lead to health hazards to human being and animals.

ii. Pollutant- The common pollutant are nitrate and polychlorinated compounds. The nitrate contamination in milk occurs through additives from drinking water, pasture and forages fertilized with nitrate compound. Consumption of nitrate polluted milk may lead to genotoxicity, carcinogenicity, local dehydration, irritation and blood dieresis etc.

iii. Heavy metals and radio nuclides- Heavy metals contamination in milk occurs when dairy animals ingest polluted feeds, pharmaceutical medicine, fodder drinking water, food additives and other materials during transportation processing and packaging of milk and milk products. Various surveys conducted by different workers revealed that Indian milk content much higher levels of heavy metals than that of internationally recommended maximum residue.

Steps of Organic Dairy farming:

1. Source of feed: All feed should come from from farm itself or procure from local organic farm only. Certification standard allow a percentage of feed consume from conventional farm subjected to a maximum prescribe limit. Synthetic appetizer, synthetic growth promoter stimulants, preservatives, urea, artificial coloring agent farm animals byproducts feeds subjected to solvent extraction droppings dung or other manure, pure amino acids and genetically engineered organism or their products are strictly prohibited in feeds.

2. Breeding: All organic animals should be born and raised under organic management, if not available calves can be brought in from conventional animals up to 4 weeks fed on colostrums. Artificial insemination is permissible, embryo transfer technique and use of genetically engineered species or breeds are not allowed. Hormonal heat treatment and induced births are not allowed except for medical reason. However breeding management should be according to animal's natural behavior and good health.

3. Treatment: Natural medicines and methods including ayurvedic, homeopathy, unani medicines and acupuncture should be emphasized. Conventional vetenary medicines can be allowed when no other justifiable alternative is available, but in all such cases withdrawal period should be double the legal period.

4. Transport: Tranquilizers or stimulants should not be given to prior to or during transport. Stress of animal should be minimum during transport. Mode of transport should be appropriate for each animal, and animals should be regularly fed and given water during transport.

5. Length of conversion period: Both livestock and fodder form should be converted to organic. Animal product may be certified organic only after farm has been under conversion at least 12 months. In dairy and egg production the conversion period should be 30 days at minimum.

The National Policy on Organic Product (NPOP) - has appointed agricultural and the processed food exports development authority (APEDA) as accreditation agencies for organic livestock production in India. Evaluating agencies appointed by Accreditation agency consist of experts from outside and acceditational agency personnel. It submits an evaluation report along its recommendation to accreditation agency for consideration to particular inspection and the certification agency. Complete written health records and field records are required for certification. This verifies your practices and it is believed that keeping records of all health problems and the prevention practices can help one to graduate from a crisis treatment pattern to good prevention management.

Steps to promote organic milk:

1. Strong domestic market should be developed to protect benefits of producers.
2. Disease free zone should be established to make organic milk production easier.
3. Research and developmental programme should be organized across the country
4. To provide technical support, State agriculture Universities (SAUs), Krishi Vigyan Kendra (KVK), Central Agencies, Reputed NGOs and Agro-clinic should be established to act as service provider and Training and extension programme should be organized for farmers.
5. Approval of accreditation and certification should be kept as National level and should be affordable to small farmers.
6. Awareness of benefits about benefits of organic milk production should be developed by government agencies and NGOs to increase.

Consumption and government subsidies should be provided to those farmers who adopt organic milk production.

Limitation in Organic Livestock Production:

The people in developed countries have resources to pay for high quality organic products. Though Indian farmers are resources poor, they cannot afford to remain aloof to the changes in the global market and requirements for exporting the animal husbandry products in the liberalized economy. However, following are important limitations in adopting organic livestock production.

1. Indian agriculture has been moving from low inputs usage to increasing use of high yielding varieties, fertilizers, crossbreds, allopathic medicines, antibiotics, growth promoters, feed additives etc.
2. As majority of Indian farmers are illiterate and less informed, they do not usually keep written farm records.
3. As artificial insemination (A.I) is not allowed under organic production, the massive crossbreeding programme promoted by the government would have its own implications.
4. As the field under production of fodder crops cannot be applied any commercial fertilizers and pesticides for the previous three years, it poses enormous practical difficulties for organic milk production.
5. The farm area required per animal under organic farming is higher than that of the conventional production system, due to higher green fodder requirement.
6. The yield per animal is comparatively lower than that of conventional farming. Initial low yields could deter new converts to organic farming.
7. Employing children in production, processing and marketing of the dairy products are to be discouraged in organic production.
8. The inadequate availability of ayurvedic, homeopathic and other herbal products will hamper the organic production.

9. Insufficient availability of clean drinking water, good quality feed and fodder and grazing area are other constraints in promoting organic production.
10. Inadequate information on national inspection and certification bodies in a comprehensive form can only dispel the prospective buyers or exporters.
11. Lack of training opportunities and incentives for organic production and market development.
12. The certification process is too expensive for the average Indian farmers.
13. In a bulk demand market, a farmer producing a small quantum requires social mobilization and region linking.
14. Non availability of nearby markets coupled with unreliable transport facilities will directly affect the price of the produce.
15. Low involvement of government, lack of subsidies, lack of clear cut policy, less effort to develop domestic market and lack of proper marketing network are some of the problems of organic products marketing.

These limitations to organic production can easily be overcome and converting to organic production maybe easier for Indian farmers as they use low external inputs for production.

References:

Indian Dairy Man May 2007, Organic Dairy Farming: Issues and Opportunities by Kadian K.S, Page 23-30

Indian Dairy Man-November 2008- Clean Milk Production residue and Contaminants problems and its impact on Human Health and Global Trade by De Kumar Ujjal et al, Page 61-68

Indian Farming in 2006 volume 55 no.10 ISSN: 0019- 4786 Organic Farming by Rajendra Prasad. Page 4-6

Indian Farming- ISSN: 0019-4786, January 2006, volume 55, No. 10. Page 3 Organic Farming by Rajendra Prasad

Indian Farming July 2008.Vol.58 No.4.Organic animal husbandry in India by A.M.Verma,P.K.Singh,D.Pandey.Page 29-30.

Indian Farming, 2007 Volume 57 no. 7 Organic Farming- Potential and possibilities by T.P Singh and Y.V Singh. Page 22-25

Wikipedia.org/wiki/organic-milk

* * *

7

RATIONALITY OF ANCIENT CATTLE MANAGEMENT PRACTICES IN THE ERA OF PRESENT CONTEXT

M.K.Tiwari
Agriculture Extension Officer, Zila Parishad,
Collectorate Campus, Bhilwara (Rajasthan), Pin-311001
E-mail: mkt_sept@yahoo.com

ABSTRACT

Amongst the allied activities related to agriculture, animal husbandry is another important sector in the country. Our country is rich in livestock resources. A large number of small and marginal farmers, agricultural laborers and other local poor folk depend upon livestock for gainful employment. It is an integral part of agriculture and plays an important role in State's rural economy by providing employment and stabilizing the household income. After agriculture, cattle and other livestock such as Goat and Sheep are the most important sources of livelihood in Rajasthan. According to the FAO, 70% of the poor keep livestock which are not only a source of food and income, but also a source of textiles, fertilizer, draught power, status, credit and cultural identity. The climate change, extreme climatic events and variability in monsoons, temperature, natural catastrophes etc. are already manifesting adverse impacts on a range of sectors including agriculture, food and water security, health, livestock, avian, forests and livelihoods. It is also leading to loss of Bio diversity with its longer consequences of loss of ancient knowledge and information systems on cattle production and management, ethno veterinary wisdom, early warning systems such as weather forecasting, and other natural resources management.

Introduction:

Ancient wisdom and information systems have a prominent role for livestock development and management and the regarding information in gaps from scientific agencies and department would have been badly advised only to rely on modern development and management practices. Ancient knowledge is sensitive and comprehensive and can be adapted to also changing climatic conditions. Indigenous nomadic pastoralists have developed a variety of livestock breeds based on traditional knowledge and have customarily grazed Cattle, and small ruminants on communal lands and in forests.

Local fodder species, local breeds of livestock provide the necessary resilience. Ecological agriculture and indigenous systems with livestock as an integrated component have the potential to mitigate some of the adverse effects of climate change. Farmers need only to take small steps to adapt management practices to benefit fully, while also contributing climate change mitigation (Nitya, 2008)

It is experienced that livestock development technology and research is biased towards commercial farm management practices leaving out the large number of marginalized livestock owners. The identified key areas of research on which focus is urgently required and will project the likely impact of such research on livestock growth, which would meet community need on increasing nutritional and food security and employment.

Practical Utility:

Cattle management as understood as specialized dairy farming is more an exception than a rule with the Region. The Indigenous patterns of animal production among the small farmers, even though only a secondary occupation to the subsistence farming of agriculture crops, has contributed for rural agriculture development by bettering the prospects of self sufficiency through self reliance and lessening dependence on outside resources. The emphasis on the studies of ancient and traditional cattle production and management has gained currency world over and a new era of practitioners in veterinary medicine and sustainable dairy farming have emerged. This approach has received support because of hazardous effects of environmental degradation. Therefore, the concern is to improve the quality of life rather increasing quantity. So, it necessitates fresh thinking to draw the useful from the ancient practices and combining it with present one (Traditional & Modern) in a manner which meets the needs of the mankind on sustained basis.

Ancient animal husbandry knowledge systems and practices are still relevant today. The emerging problems of modern cattle feeding management is defamation and chemical manipulation, it may only be possible in large commercial farms, but in developing countries like India, where most of the livestock are reared in villages in small herd, it may not be possible to follow the above methods. So this research paper may also be important in terms of cattle feeding management practices.

Ancient sages identified men as an integral part of Nature and stressed the importance of maintaining complete harmony with all the living and non-living components of earth. The ancient masters advocated leading a healthy and happy life which later evolved as customs and value oriented traditions. Unfortunately most of these great traditions are either lost or fast disappearing in this age of technological civilization. The disappearance and disregard for many of these time tested environmental conservation/ preservation oriented traditions had begun to show its detrimental effect on the life supporting system. There is now an urgent need to reinvoke and reestablish these useful traditions in the original form or in the modified form to suit the changed condition of modern living.

Therefore, the practical utility of ancient husbandry knowledge system can not be simply eluded. Unfortunately it has been dismissed as obsolete or belonging to ancient period and non-scientific. However, in recent times, its importance in the development of appropriate technologies which can sustain the process of development is being recognized. Many of these ancient and traditional practices are still prevalent in several parts of the country. It indicates that they have scientific rationality and practical utility. Keeping in view a study was conducted with the objective to document the ancient cattle management practices and its rationality in modern context. This research paper covers a wide variety of published authentic literature based on ancient Indian scriptures. The information gathered has been screened and synthesized in a manner that its objective is met. The attempt is made to justify the ancient wisdom in logical manner. Further it is to mention here that the opinions were sought from the subject specialists to prove scientific validity of the ancient concepts.

The status of Cow in Ancient India:

The Cow has deservedly occupied a pivotal place in the social, economical, political and spiritual spheres of Indian civilization since the very beginning of civilization.

The Vedas – Veritable repository of ancient Indian knowledge, are replete with references of the various aspects of cow in those times. These aspects range from the obvious milk and herding to the use of cow in bartering and extend to the atonement f sins by donation of cows. The pointed references made in the Vedas and other epics especially the childhood stories of Gopala- Krishna bear ample evidence of the fact that, in those hoary times, the economy of the agrarian society revolved around the cow and its products.

The fifth Veda- Ayurveda bears testimony to the role of cow products like milk, curds, butter milk, butter, Ghee, Gomutra, G-vara, Cow Horns and even body hair of the cow in promotion and preservation of health as well as amelioration of diseases.

Feeding of cattle was regarded as a sacred duty by the people in ancient India. Even in modern age a lot a contribution is made by the people to different Gosalas for feeding animals. Offering chapati (go-grass) and stable garbage to the cows is common practice in India. According to Padma Purana "whoever offers land for pasture of cows can never be separated from heaven, in other words, he is sure to dwell in heaven permanently. Added to this, he also acquires the merit of having donated a cow.[44]

The research paper describes the practices of the ancient cattle keepers which were developed under the perfect natural system where their animals survived on community grazing lands and also agricultural and domestic wastes. The cattle husbandry information are gathered and arranged in terms of management practices such as breeding, feeding, housing, health care, handling of milk and milk products and allied operations, etc.

1. Breeding:

People realised the economic importance of the domestic animal and as such various animal including cattle were tamed and reared for various economic purposes. However, cattle breeding appeared to be one of the important aspects of the animal husbandry practices in ancient India.

1.1 Typology of cattle:

The Vedic texts refer to cows of different colours and varieties,[1] such as, rohini (red), syemi (white), krsna (black), brown and yellow. Some were of similar colour and some of different ones. Kuta was the hornless cow, slona was the lame cow, banda was the maimed cow, and kana was the one-eyed cow.[2] Satapatha Brahman[3] mentions that the animals belonging to the family of cow like Rsabha (special bull), Babhru gau (brown ox). Prsan gau (pie-bald bullock), vasa (sterile cow), sitiprstha gau (white-blacked bullock), vasa-prsni (spotted sterile cow), syama gau (dark-grey bull), dhenu (milch cow), syeta anadvan (reddish-white draught bullock) anadvan (draught bullock), yama gau (twin bullocks), dvirupa gau (bicoloured bullock), krsna parimurni paryarini-gau (black, decripit diseased bullock), napumsaka gau (castrated bull), pasthavad (ox), uksa (bull), calves of different ages, such as tryavi (of 18 months) dityavad (two years old), pancavi or 2 and half years old, trivatsa or three years of age and turyavad of four year old, Krsna-sukla vatsa (a black cow which has a white calf)s and anaduhi vahala (yoke trained cow). Kautilya[4] prescribed on of the rules of superintendents of cow to maintain a record (register) classifying the cattle as male calves, steers, tameable cows, draught oxen, yoke and breeding bulls, fit for slaughter, buffaloes, female calves, heifers, pregnant cows, milch cows, barren cattle etc.

1.2 The Selection of Cow:

"Domesticated animals show certain characteristics which can be statistically recorded with in the overall range of variation. The environments of domestic animals and man brings about the selection of such combinations of characteristics. Selected breeds created by men through selection of animals according to definite breeding objectives and through sexual isolation show a narrow range of variation".[5]

The selection of cow for higher productivity was a specialized subject even in the days of Matsya Purana and Brihat Samhita. This gives the instruction for the selection of a cow. According to Matsya Purana[6], "first of all the cow is to be examined; there should be a cow of gentle temper free from ailment and disease, strong, of nice colour, having beautiful hoofs and horns, well-built, of middle height, giving good milk, having curls on the body (especially curls turning to the left on the right side and turning to the right on the left side) having all the lucky signs, with extensive thighs, red lips, neck and tongue, with eyes clear and beautiful (not red or having many hair) and hoofs large, having eyes of the lustre of Vaidurya, with lovely eye corners, having seven and seven teeth and bright palate, with lovely sides and thighs with six parts elevated, five parts level and eight parts capacious and wide. A cow having these qualifications is said to have auspicious signs".

Matsya said: - "The following six parts of the cow's body should be elevated viz. - chest, back, head, belly, loins".[7]

"A cow with the following level parts of the body is said to be a fine one, viz. - ears, eyes, forehead, and the following eight parts should capacious viz., - tail, dewlap, udders, thighs; and extensive head and neck are also desirable".[8]

Regarding, the features of cow's Parasara[9] observed as under:

1. Cows whose eyes are dim and with tears and disagreeable to look at and whose eyes resemble those of the mouse, will bring on misery. Cows whose horns are moving and flat, whose bodies are rough and whose color resembles that of the ass, will also bring on evil.
2. Cows having ten, seven or four teeth and those whose heads are hanging and hairless, whose backs are bent or depressed, necks short and thick, hip resembling a grain of barley in shape, whose hoofs are broken likely to bring evil.
3. The cow's with long, red, black tongues with small ankles, with a large hump on the back, with thin slider bodies and with either defective or excessive organs, will bring on evil.

1.3 The Selection of Bull/Oxen:

All the scientific breeding depended on the sturdy and physically fit bull, so at the time of performance of the ceremony of setting it at liberty, it was ensured that the bull possessed "all marks" i.e. not deficient in any limbs. Vishnu[10] says that after washing and decorating the bull, it was brought alongwith four young well bedecked cows. Apart from other religious and purificatory acts in the said cow, the mantra of "the father of the calves" was pronounced in the right ear of the bull.[11] It shows that for healthy breeding it was essential for the cow to remain in the company of a sturdy bull. That is why it was considered as the husband of the cow and the father of the young calves.[12]

Often a physically fit bull[13] was set at liberty for breeding purposes. It is laid down that it should be imposed on the person who set it at liberty for the damage done by it.[14] In the Matsya Purana[15] it is laid down that this brahmani bull must have elevated shoulders and hump, a shoft and straight tail, tender cheeks, broad back, shining eyes, sharp horns, thick and long hair on the tail, eighteen healthy teeth and eyes like Atallika flowers.

In selecting a breeding bull similar advice is given by Varahamihira. Kautilya's time, the administrative authorities appeared to be fully aware of its importance. The King was enjoined to preserve the breed of the cattle in the country and a government officer called superintendent of cattle was entrusted with the exclusive task of supervising the livestock in the country, keeping census and ensuring their proper breeding and management.[16] The selection of cattle for higher productivity was a specialized subject even in the days of Brihat Samhita which gives instructions regarding good features of cattle and the selection of a bull.

As the bull was likely to sire many of the calves born in the area, and as it was in a sense public property, its careful choice was a form of selective, and of value to the economy of the locality.[17] The gifting away of the bull to a Brahmana was a meritorious deed as the giver saw in his own cow-pen growth and increase of his cows.[18] However, the gift of own ox means the giving away of ten cows and is equal to the gift of the whole earth. Therefore, no other animal than an ox deserves veneration.[19] Moreover, the nostrils of an ox should not be pierced for three or four days after it has become strong-limbed and should never be pierced that time as long as he is weak.[20] Professionals should make the nose-pins, twelve angulas in length from the khadira or sheesham wood and should have pair or three of them.[21]

In Mahabharata cattle breeding was a regular profession. Sahadeva acts as a cowherd under Virata. "Under me" Cows multiply in number in a short time; nor does any disease appear among them. I know the marks of oxen which are fit be prized and by smelling whose urine even a barren woman brings forth a child".[22] Livestock-breeding and hunting retained important place in ancient Indian economy, especially in areas with favourable climatic conditions and a lower level of socio-economic development than in the Ganges Valley.

Ashoka's edicts show that the state paid much attention to promoting livestock-breeding along with such major officials as the Dharmamahamatras, they mention special officials supervising pastures-Vrajabhumikas.

There were certain restrictions to castrating bulls-emperor Ashoka issued an order that:

"........Athamipakhaye
dasaye Pumnadasaye tisaye puna
vasume tisa catumasisu sudivasaye
gone no nilakhilaviye ajake edake."[23]

That is to say, a bull, a goat or a ram must not be castrated on the 8th, 10th, 15th, and 13th day of each fortnight neither on the Punarvasu day, on a festival day and in every fourth month of the year.

Brahmani bulls were objects of special attention on certain festive occasions. They were marked on the right flank with a discus and on the left flank with a trident. In the Brsotsanga ceremony which was to take place on the day of the full moon in month of Kartika or Asvina, the bull was set at liberty. It was of marked as above and then washed, adorned and brought near with four young cows which were also washed and decorated. To the right ear of the bull, the mantra "The father of calves" was pronounced and also the mantra "This young bull I give you as husband" was uttered into the ears of the cows. Visnu Puranas directs that the bull must be the offspring of a milch cow having young ones living. It must not be deficient in any limb and it must be one who protects the herd. In Matsyapurana, it was instruction that the bull must have elevated shoulders and hump, a shoft and straight tail, tender cheeks, broad back, shining eyes, sharp horns, thick hair on the tail and eighteen nice teeth. Further, the bull must be well-built, roaring like the thunder clouds, high in stature and walking like an infuriated elephant.

The bulls so set at liberty were public property. They were the breeding bulls and that is why the ancients were so particular as to their physical fitness. The Arthasastra says that a herd of ten heads of either cows or buffaloes shall contain four male animals (and in Sukadum Nask of Dinkard, Book III, we find particulars about the time of allowing admission of the bull to the female). Similarly a scholar revealed that Cattle should be grouped in herds of ten each of similar colour, while they were being grazed. A certain proportion of cattle must be males; thus in every herd of ten cows or buffaloes there must be four males."[24]

1.4 Rationale of Breeding:

The second point of Brihat Samhita point out that the symptoms of cow indicate ill health. Obviously, unhealthy cow will not be profitable to the owner or may cause even death of the cow. Both the situations will reduce the profit from cow keeping. Therefore, ancient writers were knowledgeable about the health and disease symptoms of the cows. Similarly, cows with loose and flate horns as these are genetically determined traits. In ancient India the main locomotive power was bullocks which used to drive carts, pull plough, and take out water from deep bells. Such bullocks were tied in yokes which used to rest on their soldiers and the bullocks with flat and loose horns which were reluctant to perform the work used to take out their heads from the rope tied to the yokes easily which otherwise would not pull their heads had they possessed tight and upright horns. This fact was known to the ancient Indians and is also suitable while selecting a draught bullock. The ancient Indians wrote this feature of the bullock in the form of evil so that people may not commit mistake in purchasing a bullock.

However, as evidenced from the ancient literature even now a days the best males were used as breeding bull and others were castrated to be used as oxen. In present days also only best males are permitted in the herd to produce future offspring's and the rest are culled for various purposes.

In ancient times the best young breeding males were physically examined for any deviation from the breed characteristics, identifiable disease occurrence; old age, poor libido and if these characters were present, the animals were either preferred for oxen or were sold for other purposes, except breeding. The same practices are adopted now days even to physically identify the best breeding males.

As far as the scientific inference can be drawn with literature is concerned, no inference can be drawn with regard to size and shape of the horn and its relationship with milk production traits. The reason for not choosing black, brown or dark coloured oxen may be the fact that as per the scientific laws/principles the black bodies will absorb greater amount of solar radiation as compared the white/light coloured animals and thus, the efficiency of darker animals will be adversely affected in the field during sunny hours when most of the agricultural operations are performed in animal driven systems.

Similar to ancient practice during Asoka's reign, a separate animal husbandry & dairying department under government control (State/Central) is responsible for organizing all animal husbandry activities for betterment and improvement of livestock wealth of nation.

2. Feeding:

Ancient India has provided the most concrete identification of the art of feeding for the economic benefit of man. In the vedic period, people were primarily a pastoral and naturally cows and bullocks were their most valued possessions. They were always in search of new pastures, And as mentioned in the Rigveda[1] the herdsmen were entrusted with the task of taking the cattle to the pastures and forests for grazing and bring them back in the evening safely. As observed by the Rigvedic Seer Maitravaruni Vasistha, Agni looks upon the people of the world as a herdsman watches his cattle.[2] The seer Barhaspatya Bhardvaja addresses five hymns of the Rigveda[3] to Pusan who is said to be holding a tuft of a leather strap as the controller of cattle (*Pasu–Sadhani*)[4] playing the role of a herdsman. The seer Kanva Ghaura prays Pusan for giving wide pastures.[5] The seer Gathina Visvamitra praises Indra as a good herdsman who nourishes the cows and feeds them with grass, and guards them.[6]

The Rigvedic seer Mathita yamayana[7] speaks of the herdsman going to the grazing ground with kine along various paths and returning home with them to the delight of all. The seer *Sabara Kaksivata*[8] entreats Rudra for making the blowing wind healing for the kine so that they may eat herbage full of vigorous juice and drink waters rich in life and fatness. The seer *Budha Saumya*[9] shows how cattle were given pure drinking water poured into wooden troughs and bound with straps so that they were carried easily from the wells to the cow-stalls.

2.1 Pasture Grazing of Cattle:

Cows were fed on grains in addition to the grass on pastures. The vedic texts indicate that grazing was the easiest and most convenient form of feeding the animals. There were uncultivated tracts of land around the villages. The animals went there for grazing. It was an easy affair event to maintain larger number of them on pastures.[10] *Gocara* was a common term[11] to denote the pasture, which meant the place for the movement of cattle. *Vraja* also denoted the same, i.e., the place where the cattle went for grazing. *Vrajapati*[12] was incharge of the *Vraja* which was held in common by all the villagers or a number of villages. The *Vrajapati* led the villagers to the tribe to pastures sometimes for capturing the grazing animals of the enemies. The vedic texts describe the process of the driving of animals to the pastures to provide them with fresh grass.[13]

Pastures were very useful for cattle rearing and their possession was desired so that the animals might graze with ease and drink limpid water while wandering.[14] Salt was also served to them as it was essential for seasoning their food or drink. Even saline soil was considered good for them.[15] The lowing of the cow was noticed as a signal that her need was to be fulfilled by the owner. More often than not it indicated hunger and the owner was to give her more fodder for appeasing it.[16] The animals lived on grass and leaves of plants, hence Soma, the lord of plants, was worshipped as their deity.[17]

Manu gives some rules for the allocation of land for pastoral purposes and some instructions for the guidance of the keepers of animals. "A strip of land to the extent of 400 cubits or to the width of three large sticks at the outskirt of a village should be set apart for pasturage; a ground of thrice as much measure should be kept reserved for the purpose at the outskirt of a town or city."[18] Every village was again provided with common pastures lands and wood lands. Common rights in forestry and pasture were very important; and in all royal grants of villages, special provisions were always made for them. We find Manu enjoining that "on all sides of a village, a space one hundred *dhanus* or three *samya* throws (in breadth) shall be reserved for pasture and thrice that space round a town.[19] Regarding for remuneration of herdsmen, Manu says, "for tending hundred cows, a heifer shall be given to the herdsman as wages every year; for tending two hundred cows, a milch cow; and he shall be allowed to milk all the cows every eighth day".[20]

Infact, cattle rearing was one of their main occupations during Lord Mahavira and his times. The pasture grounds were known as *gochara*. During the daytime, cattle were taken out to the grazing land by the cowherds (*Gopalakas*) and returned by the evening to the people.[21]

2.2 Rules and Regulations for grazing of Pasture:

Grazing areas and grass lands, *Gochar Bhumi* were kept reserved in abundance everywhere. People used to donate their lands to *Goshalas* or even to established *Goshalas* on any auspicious occasion or to keep alive the memory of a departed family member and this was considered as a most respectful deed in the society. "Certain rules were laid down by the Kautilya for the grazing of cattle the cowherds were to look to the safety of the cattle. They were to graze the herds in forests which were severally allotted as pasture grounds for various seasons and from which thieves, tigers and other molesting beasts are driven away by hunters aided by their hounds. Sounding bells were to be attached to the necks of timid cattle, with a view to scare out snakes and tigers, and as a definite means of knowing the where abouts of herds. Cattle should be grouped in herds of ten each of similar colour, while they were being grazed".[22]

Krishi-Parasara contains elaborate rules of management of cattle. Regarding taking out cows for grazing rules are as follows :

The *three Purvas* (Purva Phalguni, Purva Shara, Purva Bhadrapatha), *Dhanistha*, *Jyestha*, *Krittika*, *Mrigasira*, and *Satabhisha* are the constellations that are always auspicious for letting the cattle in and out.[23]

One should not take the cattle out or introduce them into the shed under the constellations– the *three Uttaras* (Uttara Phalguni, Uttara Shara, Uttara Bhadrapatha), *Rohini*, *Pusya*, *Sravana*, *Hasta* and *Chitra*, as well as on the day preceding the new-moon day and on the fourteenth and eighth days of the lunar month. His cattle and other grazing animals perish who transgresses this injuction.[24]

If the cattle are taken out on Saturdays, Sundays or Tuesday (for grazing for the first time) then calamity would befall them and if they are led into the shed on those very days, then it would befall the owner of the house.[25] This practice seems to have no relevance in modern age.

Bullocks nourished on sugarcane (*saccharum officinarum* L.) fodder, barley (*Hordeum Vulgare* L.), wheat (*Triticum* sp.), and such other food and left for grazing every morning and evening are never weakened.[26] The following things brought in the vicinity of a cow pen ruin the cattle : (i) bronze metal; (ii) water kept in a bronze pot; (iii) hot scum of boiled rice or of any other grain; (iv) water that was used to wash fish; and (v) ginning of cotton (*Gossypium* sp.).[27]

2.3 Specific Ration of the Cattle feeding:

After harvesting the crops, cattle grazed on cultivated fields and also on current fallows. The weeds on cultivated lands, plants growing up from the seeds falling before harvest, the stubble and the grasses on field borders and along water channels were also available to cattle in olden days as now. Fodder crops were cultivated and made into silage – an old process in India as the word *Sujavas*[28] is mentioned in Rigveda. It indicates that the farmers were preserving fodder in the form of hay for their stock.

As the ordinary green grass of the forests was not sufficiently nutritious (to meet the nutritional requirements of animals) and wholesome, it was, therefore, supplemented by healthy food, good fodder, sufficient quantity of water, crop residues and nutrient oil cakes, etc. Thus Kautilya[29] lays down the detailed rules of quantity and a variety of food to be given to different classes of animals such as the bull, cows, horse, elephant, camels, asses and mule etc. "For bulls which are provided with nose-strings, and which equal horse in speed and in carrying loads, half a *bhara* of meadow grass (*yavasa*), twice the above quantity of ordinary grass (*trna*), one tula (100 *palas*) of oil cakes, 10 adhakas of bran, 5 palas of salt (*mukhalavanam*), one kudumba of oil for rubbing over the nose (*nasya*), 1 prastha of drink (*pana*), one tula of pulp of fruits, one *adhaka* of curds, one drona of barley or cooked masha (*Phraseolus Radiatus*), one drona of milk; or half an *adhaka* of sura (*liquor*), one prastha of oil or ghee (*sneha*), ten palas of sugar or jaggery, 1 pala of the fruit of *srngibera* (ginger) may be substituted for milk (pratipana). The same commodities less by one quarter each will form the diet for mules, cows and asses and twice the quantity for buffaloes and camels".

Apart from this the provision of subsistence to cattle, Kautilya prescribed as follows :

- Draught oxen would be provided with subsistence in proportion to the duration of service rendered.
- Milch cow would be provided with subsistence in proportion to the milk obtained from them.
- All cattle would be provided fodder and water abundantly.

"One should keep the milch cow fed on black gram leaves, or sugarcane or arjuna (leaves), well-nourished, with for nipples, of brown or black colour, having her-calf living and of the same colour, with her horns upwards, of mild nature and having thick milk. The milk of such cow alone, either boiled or unboiled, added with sugar, honey and ghee is an excellent aphrodisiac."[30] The value of leaves and sugarcane for keeping healthy nipples could not be established.

2.4 Rationale of Feeding:

The modern knowledge of animal nutrition is due to research on nutrition and dietary variables, the findings of feeding trials conducted by scientists and students and the practical feeding practices being followed by the farmers from ancient time. However, there has been modifications or revisions of the feeding practices depending upon type and purpose of animal rear, improvement of their genetic make up, availability of various feed stuffs, understanding and quantification of nutritional requirements, form of presentation of feed and economy of production, etc. though feeding is an art, but feeding practices differ in various parts of India and even in different zones/districts in a state.

It was realised in the modern era after chemical analysis of the pasture grasses that all the nutrients required by the animals were not present in sufficient quantity so as to meet out the physiological requirements for growth, milk production and pregnancy of the animals with improved genetic make up for production. The analysis of leguminous fodder crops revealed that these crops are the better sources of proteins and minerals such as calcium. Sole feeding of such leguminous fodders support 8 to 10 kg. of milk production but that is associated with loss in body weight. Moreover, leguminous fodders are deficient in P and, therefore, supplementation of brans are recommended. In the beginning, the feeding of mixed (leguminous & non-leguminous) fodder crops was advocated to meet energy and protein ratio. In earlier time, sufficient land was available for the cultivation of fodder crops.

The precautionary measures had been prescribed in the 'Arthasastra' in order to stop reckless grazing in pastures, which were basically required for the maintenance of livestock and pasture land. Though, in modern era development of pasture land can not be encouraged due to increasing human pressure and decrease in effective land area, but grazing is still in practice in different parts of the country depending upon the availability of barren lands, production status of the animal and economy involved. Bulls would not be punished if they are found straying any where either in the fields or on roads. Kautilya's guidelines as a precautionary measure are relevant even today.

The feeds described in Krisi Parasara are suggested when fed in proper proportion can provide major nutrient requirements are working bullocks: (i) & (ii) bronze metal keeping for long period may be harmful due to high content of copper salt by oxidation, (iii) scum of boiled rice or any grain provides good medium for the growth of microorganism including pathogens. Thus, it may be a source of infection. Similarly washings of fish will also have similar harmful effects, (iv) ginning of cotton if consume the large quantity may cause chocking leading to digestive disorder.

Although in most of the part of the country the green fodder is available in plenty in monsoon and winter season. The fodder can be diverted to meet out the requirement of cattle in lean season when there scarcity exist, by preserving them in the form of silage or hay as was practiced in ancient time.

The specific ration described by Kautilya that the conversion of these quantities of feed stuffs into modern unit, total feed amounts is much more than actual intake (18 kg., assuming 600 kg. body weight and 3% dry matter intake). Moreover, supply of such a variety of feed stuff is costly affair and may not be feasible in modern age when economy of the production is important.

3. Housing:

A full hymn is devoted to the cow stable, cattle were protected behind walled enclosures and watchdogs stationed to guard them. Nevertheless, the thieves got in for the days could be silenced by appropriate incantations, and cattle could be stolen. Yajurveda mentions such an instance, thus: 'Like a cattle thief getting over all obstruction comes in'.[1] And, in the famous *Prasvapin* Upanishad hymn (7.55) the Rigvedic seer Maitravarumi Vasistha actually spells out the mantras meant to be the incantations for lulling the watch dogs and inmates of the house to sleep; one of the mantra reads thus: 'May the mother sleep, let the father sleep, let the dog sleep, let the son-in-law sleep, let the kindred sleep, let the people stationed around sleep.....'.[2]

It is stated in the Agni Purana that a person by endowing houses, mansions, public halls and cow-sheds becomes sin-free and goes to heaven.[3]

Fumigation of a cow-shed with the vapours of visa plants tend to neutralise the evil effects of malignant stars thereon, and make the cattle strong in limbs. A fumigation of the cow-shed with the vapours of the Devadaru, Vacha, Mamsi, Guggulu, Asafoetida and Mustard seeds mixed together would act as a safeguard against the breaking out of those diseases among its inmates, which are usefully ascribed to the evil effects of malignant plants. The bells to hung round the neck of a cow should be fumigated as well with the vapours of the drugs enumerated above.[3a] A person who lits a light in a temple or in the house of a Brahmin or cow-shed or in a junction would become a king and all deities praise him.[3b]

The period of uncleanliness (quarantine) incidental to pollution of a death in a cow-shed is only for one day.[4] A knowledge of supreme Brahman, performance of obsequies in honour of one's own departed fore-fathers at *Gaya*, death in a cow-shed and residence with in the confines of the holy field of Kurukshetra are the four-fold means of salvation.[5]

Sheds were required to protect the animals from cold, heat and rain. It was realised that they should be fed properly because only grazing was not sufficient. The cattle stalls were as essential as the houses for the human beings.[6] The *Avesta* also considers the making of resting place for the animals as one of the most excellent deeds.[7] The cattle were driven to the pasture in the morning and they returned in the evening to the house of the owner. Cowpens were considered as protective shelters for them.[8] In the night they stayed in them.[9] The entrance of cattle into their sheds had become a widely used simile.[10]

Gostha meant resting place for the cattle.[11] Some sort of enclosure[12] (*Samvarana*) and roofed constructions were built for protecting the cattle resting in a *gostha*. In spite of all such precautions, the cattle suffered in stables which caused material loss to the owner; hence prayers were offered to the Gods to save them from dangers.[13] The cattle owners followed certain rules while building their stables. The ground was to be made smooth and clean, so that the cattle could sit, stand, sleep and move with comfort.[14] The provisions for adequate supply[15] of water and fodder were made. In such stables the cows could yield much milk and other animals could also be properly nourished.[16]

In cowpens the animals were allowed to rest freely and normally they were not tied to wooden or bamboo pegs (*sanku* and *drupad*[17]). The calves were tied near them, but kept a little away from their mothers, so that they could not suck milk. *Bandhana*[18] denoted a rope or any other thing for fastening the animals. *Veska*[19] and *Samnahaua*[20] denoted a long rope used for the same purpose.

Ancient people were very much practical and laid down specific rules and formula regarding construction and sanitation of the cow-stall. Thus, in the Krsisamgraha we find :

"A cattle-shed should be fifty-five cubits square, and it should never be erected when sun enters Leo, i.e. in the month of *Bhadra*".[21]

"One who makes his cattle-shed strong and keeps it clean from dung, secures a healthy growth for his animals".[22]

"Goats should never be kept in the same shed with cows; and rice-washing, fish washing, cotton, husks, hot starch, broomstick, pestle, or spoiled food should never be kept in the cow-shed".[23]

"To safe-guard against the breaking out of diseases, the cow-house should be occasionally fumigated with vapours of *devadara* (Pinus deodora), *Vacha* (Acorus calamus), *mamsi*, *guggulu* (commiphora mukul) Bhandari (a fragrant gum resin), asafoetida and mustard seeds mixed together"[24], and "a *piniyaka* tree (asafoetida) should be planted in the cow-house with a view to improve its general sanitary condition".

Housing system of cattle is given in Brihat Parasara-Samhita as cited in Vacaspatya-Kosa. The cows should be tied (to their posts) with their face towards the south or towards the north, but never towards the east or the west.[25] In the shed (or stable) in which cows, bulls or horses are stalled, a sharp iron sickle should always be kept.[26]

Parasara in Krishi-Parasara says that keeping the cow-shed clean and beautiful, the cattle not only remain happy but also become healthy. He prescribed in his book some useful suggestions for keeping the cow-shed neat and clean :

A cow-shed with five steps (roomy enough for the cattle to walk five steps ?) is said to be good for the cattle. It must not, however, be constructed at a time when the sun is passing through the sign of Leo (July-August), as it will surely destroy the cattle.[27]

The following things brought in the vicinity of a cow-pen ruin the cattle : (i) bronze metal; (ii) water kept in a bronze pot; (iii) hot scum of boiled rice (Oryza sativa L.) or of any other grain, (iv) water that was used to wash fish; and (v) ginning of cotton.[28]

Keeping a broomstick, pestle, or leftover food and tying goats in the cow-pen also surely lead to the destruction of the cattle.[29]

A house where the housewife uses cattle urine for cleaning purpose must not expect the cattle to have any appetite.[30]

If one wants one's cattle to grow well one should not even by mistake, dispose off the cow-dung (e.g., giving it to someone else) on Sunday, Tuesday, or Saturday.[31]

Cow dung should be disposed off on any other day except these three days as disposing it on Sunday, Tuesday, and Saturday is said to be harmful to the cattle.[32]

A cow-pen where a lamp is not lighted in the evening is without the blessings of Laxmi, the goddess of wealth. Locking at such a house the cattle weep.[33]

3.1 Rationale of Housing:

The use of watch dog for protection of cattle presently existing only in the areas where the animals are maintained basically on grazing in the forest or other waste land. This is a common practice for the protection of sheep.

Still in the villages the cattle yard considered as the house of Goddess. For clean milk production cleanliness of surrounding is most essentiai. Since in the ancient days the animal husbandry was a basic source of income. People believed the cattle yard as the house of Goddess of fortune and act of its cleaning and maintenance was considered as religious activity.

Fumigation of cowshed with leaves, seed and flowers of various plants is still in practice to safeguard the livestock from the actoparasites and different microbs present in the atmosphere of cattle yard. The actoparasites like ticks grow in the land and other crevice found in the cattle yard and the ancient people were conscious of spreading diseases even though the garland of dwells around neck of the animals.

The principles of housing mentioned in the Aitreya Brahmana having its relevance today also. For ideal housing it is necessary to provide proper feeding and watering space and to protect the animals from the inclement weather conditions. The floor of the house should be smooth, clean and impervious so that the animal can comfortably walk, sit, lie down and sleep. In such good and hygienic houses the cows may yield maximum to their potential if proper feeding and water arrangement is made.

The present housing system, there are two types of housing practiced – 1. Tie-barn; 2. Loose housing. The Atharvaveda refers to the loose housing system of cattle rearing. The calves are usually housed near the milch cows shed in order to facilitate the milking. The feeding trubs are at present also prepared of earthen pot for individual animal. But on large size herd presently this type of feeding arrangement is not convenient. In loose housing system presently the continuous feeding trubs are constructed.

General behaviour and feeding habits of goats & cattle is quite different and hence never maintained in the same house.

The dairy housing should be "animal oriented" to take care of their 'comfort' and also easy to manage. North-South orientation is preferred for minimising problems of radiation. If the animals are free to move, they will shift position so as to use the shadow cast by roof of shed for cooling.

In order to protect the animals from wind usually the houses constructed from east to west and animals face is towards the north and south (to the window side). The sharp sickle is necessary to be kept in the animal houses particularly where bulls and horses are tied the ropes to protect the animals from death due to strangulation. At any moment of risk, tying the rope should immediately cut with the help of sickle.

Parasara says that the nutritional impact on growth, health and production of animals can only be effective if the animals are maintained clean and in hygienic environment and are diseased free.

The size of stand space in the tie-barn is 5.5 to 6.0 feet. Possibly the author wants to specify this measurement of steps.

If the houses are clean with the help of cow urine the gases like, Ammonia will always present inside the house which possibly reduces the feed intake of the animals and their health conditions.

When the cattle shed is kept dark (no light is available) the general welfare of the animals particularly feeding and movement is affected which lowers down the productivity.

4. Health Care:

The medical science also made much progress during ancient India. The veterinary medicine was well developed just like the other branches of medical sciences which is quite evident by the earliest references about the curative art in Rigveda which ascribes divinity to various herbs and plants. In the Atharvaveda also certain herbs and metals endowed with medicinal properties were given divine attributes and were worshipped. During the vedic period the medical profession had became more or less hereditary. However, in the vedic period (C. 1800-1200 BC). Brahma via human mediators created the Vedas as an ethical guide to the world. Some of the Vedas contain the fist treatises on the medicine of man and beasts.

India can boast of a very ancient civilization in which all the sciences had developed to a very high degree of perfection and gives an insight into the scientific knowledge of Vedic times. Medical and Physiological lore is to be found in the Vedas in its primitive form especially in the Atharvaveda which contains references to diseases and their treatment through mantras and medicines.

Agni Purana reveals the real picture of practice of veterinary medicine in actual diseases of cattle during Gupta dynasty (300 A.D. to 500 A.D.). The kings of Gupta dynasty were orthodox Vaishnavas and this statement corroborates the period of Agni Purana connected with Gupta Empire.

1. "Oil prepared from a decoction of Shringavera (ginger), bala (Sida cordifolia), powdered mamsa and makshika (honey) together with Saindhava (rock salt) should be administered in diseases affecting the horns of cattle.
2. In a case of Otalgia (karnashoola), oil boiled with the essence of Manjistha (madder), Asafoetida and saindhava salt cooked with only garlic should be applied.
3. Toothache (Dantashoola) : The fine powder of bilva (Aegle marmelos), apamarga (Achyranthes aspera), dhataki (Woodfordia fruticosa), patala (Stereospermum suavealems) and Kutaja (Hollarhena antedysenterica) should be rubbed on gums.
4. Disease of buccal cavity (Mukharoga) : Drugs mentioned under the preceding malady boiled with an adequate quantity of clarified butter, should be used in diseases affecting cavity of the mouth.

5. Glossitis (Jihvaroga): The tongue should be rubbed by the aforesaid drugs mixed with saindhava salt.
6. Sore throat (Galagraha): The powder of Shunthi (Zingiber officinale), Haridra (Curcuma longa) and Daruharidra (Berberis aristata) mixed with Triphala should be applied.
7. Hrichchhoola (Angina Innocens), Basti Shoola (Renal colic), Vata roga, Kshaya (Weakness of Phthisis) : Purified butter mixed with Tripahla should be drink.
8. Diarrhoea (Atisara) : The powder of dhataki and Ativisha should be used.
9. Disease of Gastro-intestinal tract (kostharoga) and Diseases of the dhatus (i.e. blood etc.) : The powder of Patha, Haridra and Daruharidra should be used.
10. Cough (kasa), Asthama (shvasa) : The powder of Shunthi and Bharngi (Clerodendrum serratum) should be applied.
11. Fracture (Bhaga) : Priyangu (Callicarpa macrophylla) mixed with salt should be applied.
12. Vata roga : The oil should be applied.
13. Pitta roga : The oil cooked with Madhuyasthi should be used.
14. Kapha roga : Trikatu (Shunthi, Maricha and Pippali) cooked in oil should be given.
15. Haemorrhage (Raktasrava) : The powder of Madhuyasthi and Pustaka should be locally dusted.
16. Wound healing (Bhagnakshatpuraka) : The fine powder of Haritala mixed in oil and Ghrita should be applied.
17. To strengthen the calf : masa (urada), gehu (wheat), milk and ghrita mixed together should be given.

Morsels composed of the severed sticks of Ashvagandha and sesamum should be given to a cow for the purpose of increasing her milk. A Piniyaka (asafoetida) tree should be planted in each cow-house, with a view to improve its general sanitary condition. Salt should be given to the cows."[1] The Puranas give certain instruction for providing protection of the animals against various diseases.

Vishnudharmottara Mahapurana[2] (500-700 A.D.) contains the ancient health care practices of cattle for treating the diseased animals.

Puskara said : Listen, O Rama, hereafter I shall briefly describe the medical treatment of cattle, which is important and is beneficial to them. (Slok. 1)

Oil, in which the pounded mass of ginger, Bala and Jatamamsi has been cooked and rock salt (Saindhava) and honey added, should be applied to the roots of the horns of the cattle; or in the same way ghee mixed with the powder of simi should be applied. Oil prepared with Manjishtha (Rubia cordifolia), Hingu and Saindhava or milk should be used in all types of karnamula (Karnashula-eareche). (Slok. 2-3)

Having ground honey, rock-salt, conch, Tagari (Tagara - Delphinium brunonianum) long pepper and Siha in the goats milk, the physician should make pills of the mass. These pills rubbed in ghee and honey make an excellent collyrium for the eyes. (Slok. 4b-5)

The powder of the roots of Bilva, Apamarga (Achyranthes aspera), Dhataki, Patala and Kutaja when rubbed into the gums (lit. roots of the teeth) removes tooth-ache. (Slok. 6)

O Rama, ghee in which the ingredients recommended for removing tooth-ache have been cooked, heals the diseases of the mouth, and rock-salt is useful for the ailments of the tongue. (Slok. 7)

Ginger, the two Haridra (haridra and daruharidra), and the three myrobalans are indicated in sore throat. And the recipe consisting of ginger, the two Haridra (haridra and daruharidra), the bark of Kutaja, Apamarga and Vidanga (Embelia ribes) mixed with salt removes the diseases of the mouth and allays the morbid heat of fever. (Slok. 8-9)

For pain in the heart and urinary bladder, for disorders of Vata and for injuries and bruise it is good to give the cattle the (powder of) three myrobalans dissolved in (method) ghee as a drink. (Slok. 10)

For curing all the diseases of the heart in cattle, O Rama, oil in which Satapushpa (Anethum sowa), Kutaja and Chitraka have been cooked should be administered to them. (Slok. 11)

In dysentery the two Haridra (haridra and daruharidra) and Patha (Cissambelos pariera) should be administered and in Anaha, Padmacharini mixed with ghee should be given. (Slok. 12)

In all leprous diseases and diseases of extremities as well as in cough and asthma, ginger and Daruharidra should be administered. (Slok. 13)

For the reunion of fractures Priyangu (Callicarpa macrophylla) mixed with salt is useful. In all types Vata disorders cattle should be administered oil in which Satapuspa has been cooked, it is a specific remedy for such troubles. Yushana (Tryushana - shunthi, marich Pippali) mixed with honey is to be given in disorders of Kapha. (Slok. 14-15)

In all types of Pitta disorders, the specific remedy is cow's ghee in which liquorice has been cooked and it should be administered when such troubles arise. (Slok. 16)

In Raktapitta juice of Sakhotaka (Streblus asper) is recommended to drink. In obstinate bleeding, the cattle should be given wheat-flour, Masa and mustard – all whipped in milk or water and mixed with jaggery to drink, as it is good in such cases. (Slok. 17-18)

The ointment prepared from Sesamum, Ambhakaruha (lotus), yellow orpiment and ghee is excellent for applying on the open wounds of the cattle. (Slok. 19)

Ailing calves should be made to drink Patha stirred in buttermilk or turmeric dissolved in milk for alleviating their suffering. (Slok. 20)

Masa, (Phaseolus mungo) Sesamum and wheat (ground) and made into balls in ghee and milk of animals and mixed with salt is very nourishing to the calves and strength-giving to the bulls. (Slok. 21)

Fumigation with incense prepared from Devadaru (Cidrus devodara), Vacha (Acorus calamus), Jatamansi (Nardostachys, Jatamansi), Guggulu (Commiphoara mukul), asafoetida and mustard along with Palankasha mixed with a little ghee is auspicious and wards off the evil influence of all planets. A bell fumigated with the same incense should also be tied to (the neck of) the cattle. (Slok. 22-23)

Ashvagandha (Withenia ashvagandha), Sesamum and Chukra (Garcinia pedunculata) is useful for clearing the bladder. Buttermilk mixed with Ashvagandha, and Sesamum are also useful for the bladder, and with this prescription, scions of Bhrigu race, the milk of the cow is increased. (Slok. 24)

O Rama, oil-cake is considered to be an Rasayana for the cattle. Wet grass and cold-water for drinking should be eschewed from their feed. (Slok. 25)

At night lamps should be lit in cattle-shed. (Slok. 27)

As a prophylactic measure salt should be regularly given to cattle after every fortnight, for it prevents constipation, colic and want of appetite. It is also beneficial to sheep and goats. (Slok. 28)

4.1 Rationale of Health Care:

There has been world wide rethinking over the last decade about animal health care. It is well recognised that productive animals must maintain excellent health if their performance is to be optimised. The ancient health care practices are very conducive in modern age because it helps in achieving the sustainable livestock production. Unfortunately most of the Veterinarian were not able to provide any comment regarding scientific base of the ancient wisdom especially ayurvedic medicines (ancient plants). Therefore, it was not possible to give the justification of ancient health care practices by the end of this sub-chapter as a separate part.

5. Handling of Milk & Milk Products:

Management practices of cattle for the production of milk-yield and preparation of a large number of substances formed the dairy activities in ancient India. "The early account of dairy farming in India though noticed from vedic-chalcolithic period, its existence is easily presumed from the extensive cattle culture from the neolithic period onwards"[1]. Progress and development in dairy farming by proper management of cattle with the supply of various nutritious feeds is evident. According to Ayurveda, "milk of a black cow alleviates *vayu* and it is excellent for the patient. The milk of white cow aggravates *kapha*. The milk of red cow aggravates *vayu*. The milk of yellow cow alleviates *pitta* and it is very useful".[2] Cows have paramount importance because they give us milk and from milk, we prepare a large number of substances:

1. Prati-duh/Apakva Ksira : Uncooked milk
2. Srita : Boiled milk
3. Dadhi : Sour curds (yoghurt)
4. Saras : Cream
5. Mastu : Sour cream
6. Atancana : Curdled milk (butter milk + fresh milk)
7. Navanita : Butter
8. Ghrita : Ghee (clarified butter)
9. Amiksa : Clotted curds
10. Vajina : Whey

5.1 Rationale of Handling of Milk & Milk Products:

The properties of uncooked milk referred by Astanga Samgraha is still relevant and having scientific rationality. However, producer of the milk and consumer being located far apart, the question of drinking milk fresh from the udder is not feasible.

Several properties of dadhi has been pointed out by Astanga Samgraha. Dadhi is popularly known as curd in India. Curds when become sour (after fermentation) it is not good for health, hence the warning regarding its use is highly meaningful. If it is to be taken so it should be mixed with sugar, honey, ghee, soup of green gram or goose berry.

Kautilya mentions the preservation and consumption of butter milk by human being and animals. Butter milk is a good beverage acceptable particularly in summer even in many developed countries since it contains protein, minerals and vitamins. It might also be helpful in digestion. However, butter milk is supposed to contain less of fat and thus it could be useful for those who are fat conscious and need to consume. However, his suggestion to store butter milk in bronze vessel appears to be non-scientific since butter milk is acidic in nature, therefore it will adverse react to the bronze container.

Satapatha Brahman mentions clotted curds and it has relevance even today. In present time clotted curd is being more popular in the advance society. In India, most of the dishes (sweets, vegetables & sauses) are being prepared by clotted curds, though it is costlier. Regarding whey, (watery part of curd) it contains riboflavin (vitamin B), partly protein and fat which is helpful in digestion thus it could be said that our ancient knowledge have scientific base which has been proved by modern scientists.

Mastu has similar properties as with butter milk. In addition to butter milk it also has therapeutic value for human beings. Due to acidic in nature it increases palatability as well as digestibility of the food ingredients of the human diet. Stages for milk treatment is mentioned in Yajurveda. These stages have been adopted in present with slight modifications. The procedures followed during this time are very precise, faster and asceptic condition. Villagers still use *saras* for maintaining their health.

At present butter is manufactured both indigenously as well as by modern techniques. In the modern technique bacterial contamination is reduced. Therefore, butter produced from this technique is comparatively good for the health.

Conclusion:

It may be stated that the present research paper based on observations of cattle management practices through the ages has opened new dimensions to the researches in the field of animal sciences. Talking in the specific contexts of cattle management practices it may be underlined that most of the practices traditionally being followed have deeper meaning than what meets the eye. However, we need a scientific approach to examine, evaluate and adapt them to the present day situation. This process of the re-evaluation and re-modification has always been in vogue in sustainable development of any practice including in the livestock development. If the traditional Indian wisdom in this area is to be believed, different restrictions have been made for different animals according to their functional utility and purpose for which they were based. Therefore, there is a need to examine those texts not only contextually but with a scientific and modern approach at the same time with a sensitivity to understand. Any research in this field should adopt holistic approach for the interpretation of such data to arrive at any worthwhile conclusion.

Looking for the rationale of breeding certain physical symptoms were said to be significant in ancient texts like, Brihat Samhita and Matsya Purana. To ensure healthy breeds care seems to have been taken. This genetically proven fact appears to have been known anciently a practice which is traditionally in vogue even today. In housing of cattle fumigation was a well known practice right from the ancient days. Texts like Krishi-Parasara have laid down well specified rules for healthy upkeep of cattle. Most of these practices may still be justified and therefore, relevant till date. There are certain taboos related to cattle management for which it is hard to find any rationale. For example, certain days of the week are considered inauspicious for disposal or removal of waste of cow dung. With the modern scientific temper, efforts should be made to do away with such taboos or irrational practices. At the same time the scepticism that everything is old should be discarded without proper investigation or examination is equally in act.

The rationale of consuming milk drawn immediately after, is well known traditionally. Cattle were considered to be wealth in ancient India amongst other reasons for it. Milk and milk products were one of the primary ones. It was the traditional healthful right from the Vedic age. The modern medicine treats milk to be complete food for the young and the ailing and almost essential for the healthy persons also. Thus, there is every justification in accepting the traditional wisdom in cattle breeding, feeding, housing, etc.

In course of the present study an effort was made to seek opinions of experts in relation to the ancient cattle management practices. Only a small portion of ancient cattle management practices were explained by scientists as rationale. However, in case of large number of practices the modern scientists were not in a position to comment. This could well be because of lack of information of ancient cattle management practices to the modern scientific investigations (Since the literature is largely in Sanskrit and it is scattered). At the same time, the policy makers never encouraged such investigations. Therefore, there is an urgent need to orient our researches towards this direction so as to accept, propagate and perpetuate these practices which could be most suitable for our breed and our environment. On the contrary if there is no substance in them we should be in a position to reject them all together.

Last but not least, significant point is the ayurvedic system of medicine used for the living species in general i.e. both for the human beings as well as animal world. It is needless to emphasize the eco-friendly and sustainable nature of traditional wisdom of cattle management i.e. based on easily accessible and cheaply available plants and herbs. The manyfold virtues of the Indian system of health management is being realized today more than ever before. This system, in comparison to the modern medicine has a much higher relevance at least to Indian condition. Where it has been tested, used and practiced through the ages. Therefore, the modern researches in field of veterinary sciences in India should explore the possibilities of future utilization of rich Indian flora and exploit it to its full potentials. They should be attempted with a sence of urgency as we have now to compete at an International level in this age of patent rules largely framed in favour of western world.

References:

Agni Purana, Chapter 211 verse 18.

Agni Purana, Chapter 292 verses 33 to 34.Mahabharata, Chapter 92.

Agni Purana, Chapter CCXCII, 33 and 35.

Aitareya Brahman, Panchak 3, Chapter 2, Hymn 24.

Aitareya Brahman, Panchak 5, Chapter 5, Hymn 27.

Aitareya Brahman, Panchak19, Chapter 4.

Amar. 9. 62, p. 213; 9.70; p. 215 cited from Maity, S.K. (1957). Economic Life of Northern India. The World Press Pvt. Ltd.

Atharvaveda, Mandal 1, Chapter 164, Hymn 40; Mandal 7, Chapter 73, Hymn 11; Mandal 9, Chapter 10, Hymn 20.

Atharvaveda, Mandal 12, Chapter 4, Hymn 3.

Atharvaveda, Mandal 18, Chapter 4, Hymn 33.

Atharvaveda, Mandal 19, Chapter 58, Hymn 4.

Atharvaveda, Mandal 2, Chapter 14, Hymn 2.

Atharvaveda, Mandal 3, Chapter 14, Hymn 1 to 6.

Atharvaveda, Mandal 3, Chapter 14, Hymn 1.

Atharvaveda, Mandal 3, Chapter 6, Hymn 7; Taittiriya Brahman, Kand 3, Prapathak 8, Anuvaka 9, Hymn 4; Satapatha Brahman, Kand 13, Prapathak 1, Brahman 6, Hymn 2.

Atharvaveda, Mandal 6, Chapter 115, Hymn 2 to 3.

Atharvaveda, Mandal 6, Chapter 52, Hymn 1.

Atharvaveda, Mandal 7, Chapter 53, Hymn 1.

Atharvaveda, Mandal 9, Chapter 4, Hymn 19.

Atharvaveda, Mandal 9, Chapter 4, Hymn 2 & 17.

Brihat Parasara Samhita Vol. II, Sl. 21.

Brihat Parasara Samhita Vol. II, Sl. 22.

Brihat Samhita. Features of cows and oxen, Chapter 61, verse 2 to 4.

Brihat-Parasara Samhita. Vol. II, Slok. 51 cited in Raychaudhuri (1964). Agriculture in Ancient India. ICAR, New Delhi, pp. 125.

Brihat-Parasara Samhita. Vol. II, Slok. 53.

Brihat-Parasara Samhita. Vol. II, Slok. 54.

Buch, M.A. (1945). Economic Life in Ancient India. Baroda, pp. 41 – 42.

Buch, MA (1945). Economic Life in Ancient India. Baroda, pp. 41 – 42.

Charak Samhita, Edited and Translated by Sharma, P.V. (1983) Chowkhambha Orientalia, Varanasi, vol. II Chapter II, 2. 3, 3 – 5, pp. 44.

Garuda Purana, Chapter 107 verse 21.

Garuda Purana, Chapter 82 verse 14.

Ibid., Sl. 90.

Ibid., Sl. 91.

Ibid., Sl. 92.

Ibid., Sl. 93.

Ibid., Sl. 94.

Ibid., Sl. 95.

Jain, K.C. (1974). Lord Mahavira And His Times. Motilal Banarasidass, Varanasi, pp. 282.

Krishi-Parasara, Sl. 37.

Krishi-Parasara, Sl. 38 & 39.

Krishi-Parasara, Sl. 40.

Krishi-Parasara, Sl. 86.

Krishi-Parasara, Sl. 89. Translated by Sadhale, Nalini (1999). Agri-History Bulletin No. 2 , Asian Agri-History Foundation, Secunderabad, pp. 52.

Krishi-Parasara, Sl. 90.

Krsisamgraha verse 84.

Krsisamgraha verses 86 and 87.

Krsisamgraha verses 88 and 89.

Mahabharata, (Virata Parva), Chapter 10.

Maity, S.K. (1957). Economic Life of Northern India. The World Press Pvt. Ltd.

Mandokhot, U.V. (1987) Breeding practices and selection criteria for domestication of animals. Indian Journal of History science, vol. 22 (2), pp. 127 - 135.

Manu VIII, 231.

Manu VIII, 237, cf. Yajn. II, 170 and the Agni Purana, Chapter CCLVII, verse 18.

Manu VIII. 237.

Matsya Purana, Chapter 207, verse 10.

Matsya Purana, Chapter 207, verse 11 to 12.

Matsya Purana, Chapter 207, verse 13 – 41.

Matsya Purana, Chapter 207, verse 2 to 8.

Nar. XI, 33.

National Seminar on cow in agriculture and human health- 2003. **Souvenir &**

Nitya, G. (2008) Livestock and Climate Change. **LEISA India** Vol. 10 (4). P. 20.

Rigveda, 7. 55. Cited from Kansara, N.M. (1995) Agriculture and Animal Husbandry in the Vedas, pp. 143.

Rigveda, Mandal 1, Sukta 10, Hymn 2; Atharvaveda, Mandal 5, Chapter 20, Hymn 3.

Rigveda, Mandal 1, Sukta 191, Hymn 4; Vajasaneya Samhita, Chapter 3, Hymn 21; Atharvaveda, Mandal 3, Chapter 14, Hymn 1, 56; Satapatha Brahman, Kand 11, Prapathak 8, Brahman 3, Hymn 2; Aitareya Brahman, Panchak 3, Chapter 18, Hymn 14.

Rigveda, Mandal 1, Sukta 42, Hymn 8.

Rigveda, Mandal 10, Sukta 105, Hymn 5 to 7.

Rigveda, Mandal 10, Sukta 169, Hymn 9 & 10.

Rigveda, Mandal 10, Sukta 19, Hymn 3 to 5.

Rigveda, Mandal 10, Sukta 19, Hymn 5.

Rigveda, Mandal 3, Sukta 45, Hymn 3

Rigveda, Mandal 6, Sukta 28, Hymn 7; Mandal 7, Sukta 99, Hymn 3.

Rigveda, Mandal 6, Sukta 53, Hymn 56 & 58.

Rigveda, Mandal 6, Sukta 53, Hymn 9.

Rigveda, Mandal 7, Sukta 13, Hymn 3.

Roy, B.P. (1984) The Later Vedic economy. Janaki Prakashan, Delhi, pp. 261.

Sacred Book of the East, vol. 4, p. 140.

Sashtri, M.N.D. (1967). Agni Puranam, Chowkhambha Sanskrit Series, Varanasi, Chapter CCXCII, verse 23 to 39, pp. 1075 – 1080.

Satapatha Brahman, Chapter 21, Brahman 1, Hymn 10.

Satapatha Brahman, Kand 13, Prapathak 3, Brahman 2, Hymn 3.

Satapatha Brahman, Kand 2, Prapathak 1, Brahman 1, Hymn 6; Kand 5, Prapathak 2, Brahman 1, Hymn 16.

Satapatha Brahman, Kand 3, Prapathak 8, Brahman 1, Hymn 5.

Shamasastry, R. (1929) The Superintendent of cows. Kautilya's Arthasastra, Book II, Chapter XXIV, Wesleyan Mission Press, Mysore 146 - 142.

Shamasastry, R. (1929). The Superintendent of Cows. Kautilya Arthasastra. Book II, Chapter XXIV, Weslayan Mission Press, Mysore, pp. 145.

Taittiriya Samhita, Kand 1, Prapathak 1, Anuvaka 2, Hymn 2; Satapatha Brahman, Kand 1, Prapathak 3, Brahman 3, Hymn 6; Kand 2, Prapathak 6, Brahman 1, Hymn 15.

Taittiriya Samhita, Kand 6, Prapathak 30, Anuvaka 4.

The Rajasthan Chapter of Asian Agri- History Foundation, Udaipur, Rajasthan, India.

Vajasaneya Samhita, Chapter 15, Hymn 26.

Vajasaneya Samhita, Chapter 21, Hymn 43 & 45.

VI, I, 171; II, 341. Cited in Roy, B.P. (1984). The Later Vedic Economy. Janak Prakashan, Delhi, pp. 260.

Vide P.E.V.

Vishnu Dharmottara Mahapurana, Khand 2, Chapter 43, Edited by Khemraj Srikrishnadas. Nag Publishers, pp. 203.

Vishnu Purana, Chapter LXXXVI, verse 12. cited from Kher, N.N. (1973). Agrarian and Fiscal Economy. Motilal Banarasidass, pp. 224 – 225.

Vishnu Purana, Chapter LXXXVI, verse 13.

Yajurveda, Chapter 12, Hymn 84 ab.

* * *

8

APPLICATION OF GEOGRAPHICAL INFORMATION SYSTEM (GIS) IN DAIRY INDUSTRY

DEEPAK BHAGAT* & SHWETA PRIYAMVADA**
*Assistant Professor, Department of Management, NEHU, Tura Campus
E-mail: dip19bhagat@gmail.com
** Student, Department of Geography, Cotton College, Guwahati University.

ABSTRACT

Milk is considered as an excellent food for human being especially for the vegetarians as it is the only source of first class protein for them. India with a production volume of 84.4 million tonnes in 2001- 02 became the largest producer of milk in the world. However, the biggest problem of the dairy industry is fresh milk management. As fresh milk is highly perishable the key issue is timely fresh milk delivery. To solve this logistic problem Geographical Information System (GIS) can be used. GIS will lead to a better and improved supply chain management in dairy industry. An attempt has been made in this paper to understand the application of GIS in improving the traditional supply chain of dairy industry. First the concept of GIS has been explained and then in the light of few examples its application in the dairy chain has been discussed.

Introduction:

Milk as defined by Oxford Advanced Learner's Dictionary (1998) is "the white liquid produced by female mammals as food for their young, especially that of cows or goats drunk by humans." Milk is considered as an excellent food for human being especially for the vegetarians as it is the only source of first class protein for them. It is an important or an essential item for both newly born young ones and to the expectant mothers as it contains essential elements like calcium and phosphorous along with numerous other major and minor substances.

India with a production volume of 84.4 million tonnes in 2001- 02 became the largest producer of milk in the world. In 2006-2007, it touched the magical figure of 100 million tonnes (Table 1). Milk production in India was almost stagnant during 1947-1970 with annual growth rate of merely 1%. By regular efforts by centre and state government milk production in India registered a growth rate at an average rate of 4.5% per annum during the eighth plan (Khullar, 2007). This increase in milk production is known as White Revolution. This has been possible with the introduction of Operation Flood Programme (1971-1996) by the National Dairy Development Board (NDDB) Anand. The Operation Flood is an integrated dairy development programme, which has helped 83.5 lakh farmers through 65,092 dairy cooperative societies. In 2005, the size of India's dairy sector was estimated at Rs 2300 billion, with a production of 95 metric ton (Khullar, 2007).

Table 1: Milk production in India

Year	Production (Million Tonnes)	Per Capita Availibilty (gms/day)
1991-92	55.7	178
1992-93	58.0	182
1993-94	60.6	187
1994-95	63.8	194
1995-96	66.2	197
1996-97	69.1	202

1997-98	72.1	207
1998-99	75.4	213
1999-2000	78.3	217
2000-01	80.6	220
2001-02	84.4	225
2002-03	86.2	230
2003-04	88.1	231
2004-05	92.5	233
2005-06	97.1	241
2006-07	100.9	246
2007-08	104.8	252
2008-09	108.5	258

Source: Department of Animal Husbandary, Dairy & Fishery, Ministry of Agriculture, GoI

Supply Chain Issues in Milk

Out of total milk production of the country 15% of the milk is marketed via formal channels and remaining 85% via informal channels (Hemme, 2003). Figure 1 shows milk marketing channels in formal and informal sector in India.

In the informal channels the consumer is directly connected to the creamery, milkman or farmer and they use their own container for the transportation of milk. On the other hand, in the formal sector milk is sold in plastic bags of 0.5 and 1kg.

However, the biggest problem of the dairy industry is fresh milk management. As fresh milk is highly perishable the key issue is timely fresh milk delivery. To solve this logistic problem Geographical Information System (GIS) can be used. An attempt has been made in this paper to understand the application of GIS in improving the traditional supply chain of dairy industry. First the concept of GIS has been explained and then in the light of few examples its application in the dairy chain has been discussed.

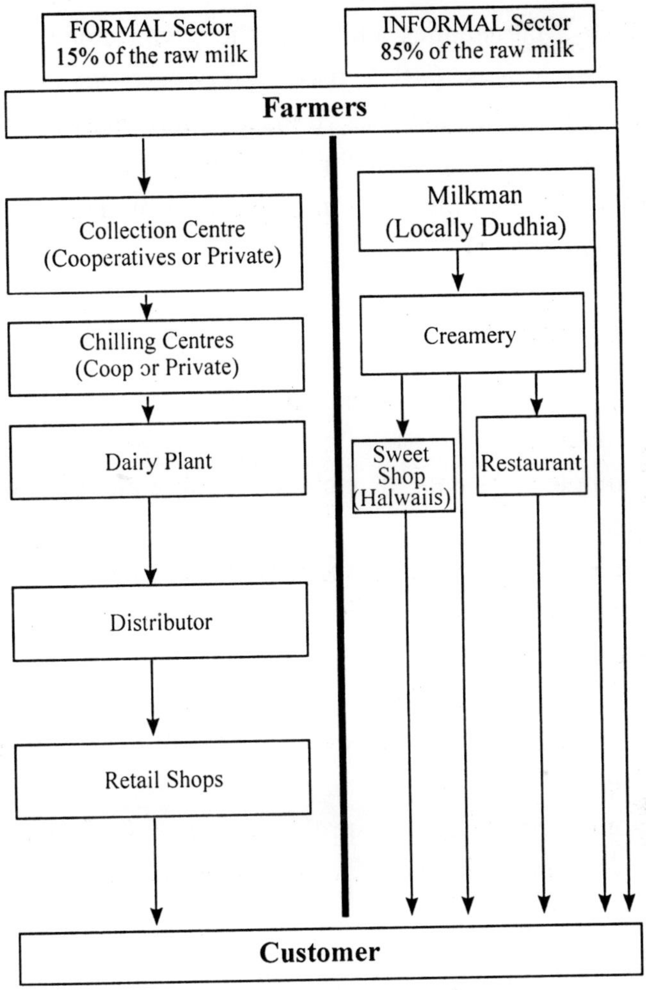

Source: Hemme et al. (2003)

Fig. 1: Milk marketing channels in formal and informal sector in India

Geographical Information System (GIS)

A Geographical Information System (GIS) is a computer based system for capturing, storing, querying, analyzing and displaying geospatial data. Geospatial data are data that describe both locations and the characteristics of spatial features such as roads, land parcels, and vegetation stands on the Earth's surface (Kang-tsung Chang, 2009).

The discipline of GIS borrows knowledge from various disciplines like Geography, Mathematics and from the advancement in the field of electronics. Now a days, GIS is used everywhere but it is widely used in- site analysis, routing analysis, transportation analysis. It is used by many other disciplines as tools for handling spatial data in geographic environment. GIS stored the maps in digital form in database in world co-ordinates (meters or feet)

Components of GIS

There are five main components of GIS-

(i) Hardware

Computer system- It includes both computer and operating system

The input devices- Digitizers or scanners etc

The output devices- High resolution printer, plotter etc

(ii) Software

GIS softwares are the programmes used for running the hardware. For example- Arc Info, Arc GIS, Map Info, Geomatica etc.

(iii) People

People are the trained professionals of GIS who describe the objective of using GIS for particular purpose and also justify the reason.

(iv) Data

Data are the inputs supplied to the system for generating information. The examples of data are- Roads, Census tracts, Water bodies, Landmarks, Bus Routes, Rail Lines, Schools, Wetlands, Traffic counts, Geologic data etc. various sources of data used in GIS are Toposheets prepared by Survey of India (SOI), Satellite Imageries and Census data etc.

(v) Infrastructure

Infrastructures are the physical, organizational, administrative, and cultural environments who support GIS operations. It includes good skills, standard data, data clearing houses, and general organizational patterns.

GIS and Dairy Supply Chain

To understand GIS application in milk supply chain three examples has been discussed. These cases are-

(i) CONAPROLE (National Cooperative of Dairy Producers of Uruguay)

CONAPROLE is one of the most important private companies in the country. It has more than 2,700 milk collecting points and more than 10 depots countrywide. More than 120 trips are done daily (about 180 in peak season) by the people involved in collecting milk. It has been possible with the help of better information system. A new GIS software named GLF-inforut II has been developed by the researchers for planning milk tanks routing to several processing centres. With the help of GLF-inforut II users can plan, control and evaluate past and future tours. Inforut II is developed with ESRI software, Arc info 8.1 and map object 2.1

CONAPROLE started the Fresh Milk Management (GLF) project for the delivery of fresh milk on time. This technology is not only helpful in delivering the map, it also calculates the payment of the dairy farmers, gives current information to the farmers and provides information to both dairy farmers and transport companies to corporate administrators.

(ii) SARSA Green (Durgapur, West Bengal)

Main objective of this project is to increase the production of milk through integrated policy of cattle cum dairy development and eliminating the role of middleman by forming the co-operatives and self supporting dairy enterprise. This project along with dairy cattle farm has other units like fodder production, hut management; cattle feed plant, dairy plant and different farming system. For this purpose the project is using GIS to get correct and precise information about the factors related with the dairy farm. GIS is helping in showing the management of land use for different purposes like cultivated land, feeding mill, house of cattle, biogas plant and irrigation process.

For the integrated dairy and farm management availability of groundwater is needed when the only source of water is groundwater. GIS techniques are used to get the information about occurrences of groundwater, which are the areas suitable for future groundwater exploration, depth and quality of water. Groundwater can be used for different purposes- drinking, irrigation and in the dairy processing plant. GIS is useful for mapping of the groundwater prospective areas and management plan of the dairy farm. All these information can be shown in one map which will help the planners in decision making.

(iii) Amul (Gujarat Cooperative Milk Marketing Federation, GCMMF)

Amul is not only a brand but it has given employment to many people (especially in rural areas) in Gujarat which other brands has failed to do. Amul products are found in over 500,000 retail outlets across India through the network of 3500 distributors. It has 47 depots with dry and cold warehouses to buffer inventory of the entire range of products. Amul collects 447, 00 liters of milk from 2.12 million farmers, converts milk into branded packaged products and sells goods of worth Rs 6crore.

Amul has applied Geographical Information System (GIS) for the delivery and management of the products on time. Amul is using GIS technique for business planning and optimization of collection process. It uses GIS at both ends of the supply chain (milk collection as well as the marketing process). Amul has also implemented GIS based sales analytics solution across various offices. For the effective and geographically monitoring of the sales performance of WDs, ADAs, retailers, and products, it has developed GIS based Decision Support System (DSS).

After analyzing and discussing the above three examples, the significance of GIS in improving the supply chain can be clearly understood. In brief the GIs can help Dairy Industry in the following way-

(i) It is helpful in finding out the shortest and least cost route for the supply of fresh milk on time.

(ii) GIS monitors the production rate, location of shops and their selling rate.

(iii) With the help of GIS software one can plan for the future tours (24-48 hrs) and also calculates and evaluates the past tours.

(iv) It calculates payments of farmer and also provides current information to them.

(v) It is helpful in providing information to corporate administrators about dairy farmers and transport companies.

Conclusion

In modern day marketing, competition is not between the firms, but between the supply chains. This becomes more critical if the supply chain if of fresh milk which is characterized by perishability. After analyzing and discussing the issue and examples, it is found that GIS can help dairy industry in finding out the shortest and least cost route for the supply of fresh milk on time, by monitors the production rate, location of shops and their selling rate, planning for the future tours (24-48 hrs) and also by calculating and evaluates the past tours. GIS can also calculate payments of farmer and also provides current information to them. It is also helpful in providing information to corporate administrators about dairy farmers and transport companies. Thus, GIS be a big support to dairy industry.

References:

Bowonder, B., Raghu Prasad, B. R., & Kotla, A. (2005). ICT application indairy industry: The e- experience of Amul. Proceedings of *IJST*, pp. 241- 265

Chang, K. (2009). Introduction to geographic information system, New Delhi: Tata McGraw- Hill

Ghose, A., Dey, J., & Dwary, P. (nil). Application of GIS on small farm and dairy management: SARSA green, Durgapur, West- Bengal.

GoI. (2010). Milk production in India. Department of Animal Husbandry, Dairying & Fisheris, Ministry of Agriulture.

Gonzalez, M. (nil). GIS and logistics tool for milk transportation in dairy industries. http://proceedings.esri.com/library/userconf/proc03/p0119.pdf

Hemme, T., Gracia, O., & Saha, A. (2003). A review of milk production in India with particular emphasis on small- scale producers, PPLPI working paper 2, FAO, Rome

http://www.gisdevelopment.net/application/agriculture/overview/agrio0010.htm

http://www.nddb.org/statistics/milkproduction.html

Khullar, D. R. (2007). *India: A comprehensive geography*, India: Kalyani Publishers

* * *

9

PROSPECTS OF DAIRY DEVELOPMENT AND INDIGENOUS DAIRY PRODUCTS IN NORTH EASTERN REGION

RAJENDRA KUMAR PANDEY
Professor & Former Head, Department of Anmal Husbandry & Dairying, In-Charge Dairy Farm, Institute of Agricultural Sciences, Banaras Hindu University,Varanasi-Uttar Pradesh-221 005

ABSTRACT

Basically India is an agricultural country and livestock sector is integrated part of agriculture. It is the backbone of Indians of India economy in the form of income, employment and foreign exchange earnings. It is estimated that dairying sector alone contributing more than 15% Gross National Income. In our country nearly 80% of people living villages 69%of them are engaged in agriculture 43% of them are cultivars having a piece of cultivable land. Remaining 26% are agricultural labourers. The livestock sectors is giving sustainability for their life, it is providing livelihood to the people living in drought areas like Rajasthan and Gujarat state etc. the tremendous increase in milk production since last 30 years. Now India becomes the largest milk producing country in the world, with annual milk production of more than 96 millions tones. Simultaneously there is an increase per capita consumption from 112gm in 1971 to 242gm in 2007. Today dairying sector estimated revenues is 1, 05,000 crores with annual growth rate of 5%, the demand for milk also increasing 4 to 8% annually. Dairy industries are one of the largest businesses in India next to agriculture. The turnover of milk and milk product in the yrs 2006 was values Rs. 227000 crore in which 55% accounted for liquid milk. This is also most promising business in India where the return will be available within a month. Today no other business can guarantee such payment.

Keywords: Prospects, dairy development, indigenous dairy products

Introduction:

The cattle rearing in India is as ancient as Aryan Civilization. Traditionally cattle have been raised by people (i) for prestige (ii) as a subsidiary occupation to crop production in rural areas (iv) for meat, fuel, cow dung and bullocks etc. To a great extent this is still the state of affairs. However, through systemic planning by the Government in the form of various Dairy Development programmes starting from Key Village Scheme (during the first three Five Year Plans), the Intensive Cattle Development Project introduced in sixties, Operation Flood I in 1970, Operation Flood II in 1982 and Operation Flood III in 1994, India has reached the present estimated milk production level of more than 90 million tones/year topping the world in terms of milk production as against production of only 17 million tones in 1950. We should be proud of this. India at present contributes 11% of the total world milk production. The estimated cattle population in India in 2004 is more than 200 millions which is 15% of the total cattle population in the whole world and India's buffalo population is 54% of the world's buffalo population.

Even though, India has made phenomenal progress in the area of dairy development, the achievement seems to be insignificant when viewed against the challenges faced by the dairy sector. Although we have huge cattle and buffalo population in the country, but due to huge human population of the country, the per capita availability of milk every day at present is about 230 gm which is quite low, although it is much higher than 125 gm in 1950. It is to be noted that New Zealand tops this list with 8.5 kg/day as per capita availability. Further, very low productivity of milch animals, huge expenditure involved in planning and implementing dairy development projects and under utilization of the large installed capacities of dairy plants are some of the challenges faced by the country in terms of its dairy development.

Because of socio-economical reasons, there are huge number of small and scattered herds in the country which is a big bottleneck for proper economic viability of dairy development. The investment in dairying has decreased from .82% of the total plan expenditure in the First Plan to .52% in the 9th Five Year Plan. This has caused serious constraints for manpower development, creation of proper communication, infrastructure, reliable data base and strengthening of extension network in rural areas. Lack of flow of latest technologies and know how from labs to the cattle farmers and development of suitable technology to suit different rural areas and its socio-economic condition have been bottleneck in dairy development. The major constraints observed in cross breeding has been low conception rate and reproductive performance through artificial insemination resulting into high incidence of repeat breeding in cross breds, indiscriminate breeding and slaughtering of high yielding animals, ill equipped artificial insemination centers and inadequate expert services. Approximately 80% of the cattle are non-descript local ones which give hardly 1-2 kg/day. Because of this the average milk production per cow in India is approximately 3 kg/day as against about 25 kg/day for average cow in U.S.A. This shows that the milk production can be increased many-folds by having more and more high breed animals in place of non-descript local animals in phases.

In addition to high genetic cattle, balanced feeding of dairy animals is the basis for successful dairy farming. Poor availability of quality feeds and fodder results in low production of milk, shortage of supply of essential nutrients resulting in poor growth and reproduction. Lack of knowledge on the part of farmers about balanced feed, quality feeds and fodder, and lack of organized markets for quality feeds and fodder add to the dimension of the problem. Another problem in speedy dairy development is lack of facilities and lack of knowledge on the part of dairy farmers about disease prevention and control of cattle.

At present not more than 20% of milk is being processed in the dairy plants under government, organized private and co-operative sectors. The absence of regular and remunerative market for milk and milk products is a major constraints encountered by farmers specially from rural areas. Lack of inadequate credit facilities to the dairy farmers and exploitation by middlemen hampers the farmers from deriving proper profit from dairying and sometimes causes discouragement and frustration to the farmers.

Dairy extension activities are always treated as secondary due to which most of the dairy farmers are not aware of various aspects of economically profitable dairy farming and the facilities available to them by various organizations. Inadequate communication, transportation and veterinary services located at far distances also contribute to the dairy development adversely.

It calls for a thorough discussion on the part of dairy scientists, policy markers, extension personnel and dairy farmers to address these constraints which may vary from place to place in degree. Use of Participatory Rural Appraisal, more intensive extension activities for effective flow of technologies and informations from lab and other relevant organizations to dairy farmers may help not only in identifying the problems faced by dairy farmers but also in finding possible suitable solutions.

The scenario of dairy development in North Eastern Region is all the more discouraging. Historically Shillong being the only important city in N.E Region, Britishers for their own consumption wanted to have high breed cattle in Shillong. They brought one bull of Holstein-Friesian breed from Orissa and another bull of Brown Swiss breed from Germany and arranged to start a multiple cross breed cattle farm in Shillong with the help of Nepalese. They encouraged Nepalese to rear cattle. The oldest cattle farm in N.E. Region was the one established in 1924 in Upper Shillong. At one time there were about 1000 cattle of mixed breeds (Holstein-Friesian, Jersey, Brown Swiss). Another cattle farm was established in Khanapara, Assam in 1926. Artificial insemination was started in Shillong in around 1965. In 1987 during Operation Flood-II, National Dairy Development

Board offered the states of N.E. Region for production of wholesome milk and its distribution at a reasonable price by collecting milk from the producers door-steps regularly paying its remuneration through Amul pattern cooperative system. Only Assam Government accepted the offer in 1987 with the establishment of West Assam Milk Producers Cooperative Union Limited (WAMCUL) in 1987. Nagaland followed 5 years later by establishing Kohima Milk Union Limited (KOMUL) in 1992. Assam has taken up an Agricultural Infrastructural Development Project through World Bank in 1998, which is likely to boost dairy development.

The number of high breed cattle of high yield variety rose to 25 lakh in 1965 in N.E. Region which was approximately 30% of the total cattle. A large number of Nepalese and Biharis were involved in this sector. But due to Nepalese and Biharis Hatao drive by local people of this region, this number has come down to only 18 lakhs at present which is only 15% of the total cattle in NE Region. This has happened in a big way in Meghalaya. Because of this, the total annual milk production in Meghalaya which steadily rose from 39,000 tones in 1970 to 64,000 tonnes in 1984, dropped down to only 46,000 tones in 1985 and could rise to only 62,000 tonnes in 2001.

The largest cooling plant was established in Naya Banglow (near Shillong) in 1962 which is dry due to Nepalese and Biharis Hatao drive in Meghalaya. Subsequently the government established another chilling plant along with pasteurizing unit in 1983 in Shillong which is still functioning. One pasteurization plant was established in Khanapara in 1962 to cater the needs of dairy farmers around Guwahati.

Amongst 8 states of N.E. Region, Assam is well ahead in terms of milk production (850/1300 thousands tones), milk processing Units (9/13), liquid milk plants (9/22), organized dairy cooperative (123/404) and the number of cattle 12000/16000 thousands). After Assam, Sikkim has done significantly well in terms of per capita availability of milk which is around 200 gm as against 80 gm being the average per capita availability of milk in N.E. region and 125 gm being per capita availability of milk in Assam as in 2000. Sikkim also has (162/402) organized Dairy Cooperative Societies and is top in the list of states in N.E. Region. Nagaland has been doing extremely well during the 4-5 years and the projected per capita availability of milk in 2004-05 is 436 gm/day for Nagaland as against 280 gm/day that of Sikkim and 90 gm/day that of N.E. Region average and 230 gm/day that of National average. Mizoram's performance is significant in terms of average milk yield/cross bred cattle which is 8.5 kg as against 8.4 that of Meghalaya, 8.00 that of Manipur and 4.5 being average of NE Region for cross breed cattle.

Looking at the very low per capita availability of milk in NE Region which is around 90 gm/day as against minimum requirement of 250 gm/day for a person as per Indian Council of Medical Research, there is a tremendous scope of dairy development in this region. It will not be out of way to mention here that the annual consumption of Amul Taja (milk) in N.E. Region is more than worth Rs 28 crores and that of other milk products from Amul is about Rs.12 crores. Further, annual consumption of milk products at sufficient high cost say Rs 25/ltt from other companies of West Bengal is also of the order of Rs.12 crores. Because of lack of availability of fresh milk and its products, people here are compelled to go for Amul and other Company's milk and milk products. If we can increase milk production to fulfill the demand of people here, they will prefer to buy pure fresh milk and its product at a rate about 70% of the rates of Amul Taja and its products which are 3-4 months old. In N.E. Region, Khoa and chhenna based milk products are very common. In addition to this Paneer and Sweet Dahi is also quite popular. Some of the main bottlenecks in dairy development in this region are (i) lack of awareness and information about dairy prospects (ii) lack of high quality breeds (iii) non-availability of quality feed and fodder (iv) lack of marketing net work (v) lack of transportation and communication (vi) geographical constraints (vii) lack of veterinary hospitals, breeding farms, artificial insemination facilities (viii) lack of awareness about the nutritional and medicinal value of milk and its products.

Looking at the acute unemployment, this is high time that youth should be encouraged to take up dairy farming/processing/management as self employment opportunity. Agriculture Universities, Indian Institute of Entrepreneurship, NIRD, SIRD and other related Institutions should give short term training programme in different aspects of dairy development. The funding agencies should also come forward in this direction. Dairy farmers should be given required know how and should be encouraged to produce high quality feeds and fodder. ICAR, KVK, KGK etc. may play an active role in this direction. Awareness camp may be organized from time to time at block level to create awareness about the nutritional and medicinal value of milk and its products and about dairy prospects in this region. NGO's, academic institutions, ICAR can contribute significantly towards this. Government should establish high pedigree breeding farms, A.I. centre, and animal health care units at block level and the dairy farmers should be encouraged to rare high breed cattle instead of non-descript cattle. They should also be given required know how about maintenance and management of high breed cattle. Government should open milk chilling centre, small unit dairy plants at block level.

I am sure that if the above mentioned measures are taken at block level, then within 5 years we can increase the production of milk and its products to at least double of what it is today in N.E. Region. We have made a humble beginning in this direction by training students of the department, in dairy farming, by developing a model dairy farm in the campus and by encouraging students to take up dairy as an opportunity for self employment. Special orientation programmes in dairy farming is necessary to create awareness in this region.

There have been operation fold program in three phases in the country so far. But their important in north eastern region has not been significant. I think north-eastern region needs a special programme Operation Flood north-east. To boost dairy activities in north-east region I take these opportunities to appeal dairy scientist, policy makers, Indian dairy board, I.C.A.R, K.V.K, K.G.K, N.I.R.D, S.I.R.D, Agricultural Universities and other concerned organization. In north-eastern region to jointly convey forward to meet the objective. This will not only boost dairy development but also give self-employment opportunities to thousand of unemployed youth of northeastern region.

Table 1: Annual Milk Production in different states of NE Region (Thousand tonnes).

Sl.No.	States	1981-82	1991-92
1.	Assam	469.00	639.0
2.	Manipur	62.00	83.0
3.	Meghalaya	58.00	50.0
4.	Nagaland	3.45	44.0
5.	Sikkim	18.00	29.0
6.	Tripura	17.50	32.0

Table 2: Estimates of Milk Production in N.E. Region 1997-98 (000 MT) (Thousand tonnes)

Sl.No.	States	Milk Production
1.	Arunachal Pradesh	44.0
2.	Assam	775.0
3.	Manipur	62.0
4.	Meghalaya	59.0
5.	Mizoram	20.0
6.	Nagaland	46.0
7.	Sikkim	35.0
8.	Tripura	47.0

Table 3: Projected Milk Production up to year 1998-99 to 2005-06 (Thousand tonnes)

Sl.No.	States	1998-99	2002-03	2004-05	2005-06
1.	Assam	815.32	940.58	1010.25	1047.00
2.	Manipur	99.73	112.72	119.84	123.56
3.	Meghalaya	54.88	56.88	57.91	58.44
4.	Nagaland	105.85	190.04	254.64	294.75
5.	Sikkim	40.49	49.45	54.65	57.45
6.	Tripura	47.21	58.75	65.54	69.23

Table 4: Share of cow, buffalo and goat milk to total milk production in NE States (000' MT) – 1993-94

Sl.No.	States	Cow	Buffalo	Goat	Total
1.	Arunachal Pradesh	21.0	-	-	21.0
2.	Assam	561.0	91.0	24.0	676.0
3.	Manipur	63.0	21.0	-	84.0
4.	Meghalaya	50.0	3.0	-	53.0
5.	Mizoram	9.0	-	-	9.0
6.	Nagaland	42.0	2.0	1.0	45.0
7.	Sikkim	NA	NA	NA	30.0
8.	Tripura	NA	NA	NA	35.0

Table 5: Per capita availability of Milk 1997-98 in NE. States (gms/day)

Sl.No.	States	1993-94	1997-98
1.	Arunachal Pradesh	62.0	118.0
2.	Assam	78.0	82.0
3.	Manipur	118.0	79.0
4.	Meghalaya	77.0	78.0
5.	Mizoram	32.0	64.0
6.	Nagaland	93.0	84.0
7.	Sikkim	186.0	191.0
8.	Tripura	33.0	40.0

Table 6: Projected per capita availability of milk in NE States in 2004-05 and 2005-06 (gms).

Sl.No.	States	2004-05 (gms)	2005-06 (gms)
1.	Assam	93.53	94.88
2.	Manipur	135.37	136.62
3.	Meghalaya	67.72	66.88
4.	Nagaland	436.87	495.01
5.	Sikkim	279.01	287.01
6.	Tripura	49.33	51.00

Note: Requirement of milk has been recommended as per the norms of 220 gms per day per capita by the ICMR.

Table 7: Average milk yield of cows, buffaloes and goats in NE states in 1991 (in litre/day).

Sl. No.	States	Cow		Buffalo	Goat
		Non-descript	Cross breed		
1.	Arunachal Pradesh	1.030	4.56	NA	NA
2.	Assam	0.912	3.74	1.770	0.227
3.	Manipur	1.400	7.96	3.16	NA
4.	Meghalaya	0.720	8.410	1.96	NA
5.	Mizoram	0.710	8.500	NA	NA
6.	Nagaland	1.710	6.760	3.160	0.380

Table 8: Total number of cattle, buffalo and goat in NE States in 1992 (in thousands).

Sl. No.	States	Cattle	Buffalo	Goat
1.	Arunachal Pradesh	324.0	5	125
2.	Assam	10118.0	959	3454
3.	Manipur	719	114	38
4.	Meghalaya	535	33	195
5.	Mizoram	59	7	21
6.	Nagaland	332	32	151
7.	Sikkim	198	2	114
8.	Tripura	949	20	429

Table 9: Annual growth rate of milking cows and buffaloes in NE States – 1992.

Sl. No.	States	Annual growth rate Of milking cows (%)	Annual growth rate of Milking buffaloes (%)
1.	Arunachal Pradesh	(-) 3.58	(-) 12.94
2.	Assam	6.75	1.031
3.	Manipur	4.56	0.00
4.	Meghalaya	0.00	0.00
5.	Mizoram	8.45	0.00
6.	Nagaland	8.76	14.87
7.	Sikkim	1.61	0.00
8.	Tripura	11.46	3.13

Table 10: Milk production status in NE states in 1998-99.

Sl.No.	States	Milk Production (000'MT)
1.	Assam	44.5
2.	Assam	798.0
3.	Manipur	65.0
4.	Meghalaya	61.0
5.	Mizoram	21.5
6.	Nagaland	47.5
7.	Sikkim	40.0
8.	Tripura	45.0

Table 11: Number of milk processing factories and liquid milk plants as on 31.3.1997 in NE States.

Sl.No.	States	No. of milk processing Factories	No. of liquid milk plant
1.	Arunachal Pradesh	Not reported	1
2.	Assam	9	9
3.	Manipur	-	1
4.	Meghalaya	-	6
5.	Mizoram	-	1
6.	Nagaland	1	1
7.	Sikkim	2	2
8.	Tripura	1	1

Table 12: Dairy Cooperatives in NE States as on 31.3.1998

Sl. No.	States	Dairy Coop. Societies, organized (No)	Farmer Members ('000)	Rural Milk Procure-ment ('000 kg/day)	Milk Marketing ('000 L/ day)	Processing Capacities ('000 L/day)
1.	Assam	123	2	3	6	60
2.	Nagaland	35	1	1	1	-
3.	Sikkim	162	5	7	7	15
4.	Tripura	84	4	1	5	10

Table 13: No. of dairy plants registered under MMPO as on June 1, 1996 in NE States (capacity '000 lit/day)

States	Cooperative	Private	Others	Total
Mizoram	1	-	-	1
Tripura	1	-	-	1
Sikkim	13	-	-	13

Table 14: Capacity of MMPO registered by Central Registering Authority as on March 1991.

States	District	Name
Sikkim	Gangtok	Sikkim Co-op. Milk Producers Union Ltd.
Tripura	East Tripura	The Tripura Dist. Milk Producers Union Ltd.

Dairy Cooperatives in North Eastern States:

1. **Assam**
   - (a) Cachar and Karimganj Dist. Milk Producers Cooperative Union Ltd., Silchar-788005.
   - (b) DSD Cooperative Milk Producers Union Ltd., Jorhat-785001
   - (c) West Assam Milk Producers Cooperative Union Ltd., Khanapara, Panjabari, Guwahati-781002. (Brand name : WAMCUL)

2. **Mizoram**
   - (a) Mizoram Coop. Milk Producers Union Ltd., C/O Directorate of Animal Husbandry and Veterinary, Govt. of Mizoram, Aizawl-796001.
   - (b) Mizoram Multi-commodity Producers Cooperative Union Ltd., Aizawl, Mizoram-796017. Brand Name : MULCO Milk.

3. **Nagaland**
   - (a) Bhola MPCS Ltd., Dimapur-797112 (Dairy Products and sale of milk).

(b) Chumukedima III MPCS, Dimapur. (Buy and sale of milk).

(c) Dairy Farms MPCS, Dimapur. (Sale of milk)

(d) Everst MPCS, Dimapur (Buy and sale of milk)

(e) Kohima Milk Union Ltd., Dimapur (Sale of milk, Brand name : KOMUL)

(f) Medziphema MPCS, Medziphema (Milk products and sale of milk)

(g) NC MPCS, Kedima, Dimapur (Milk sale).

(h) Tulovi Milk Producers Cooperative Society, Dimapur (Buy and sale of milk)

(i) Valley Milk Producers Cooperative Society, Dimapur (Buy and sale of milk).

4. Sikkim

Sikkim Cooperative Milk Producers Union Ltd., Gangtok 737102 (Milk processing).

5. Tripura

Tripura Cooperative Milk Producers Union Ltd., Agartala-799006 (Dairy Product preparation -Brand name – GOMATI)

Dairy Products Manufacture and Suppliers in NE States.

1. Assam

(i) Govt. of Assam, Directorate of Dairy Development, Khanapara, Guwahati (Products – Milk, Ghee, Paneer etc.)

(ii) Radhesham Enterprises, Fancy Bazar, Guwahati (Dealers in all dairy products).

State Govt. listing

1. Arunachal Pradesh

Area	93,743 sq.km
Population	8, 58,392

Livestock population (census 1997)

(a) Cattle 4, 52,920

(b) Buffalo 17,483

(c) Sheep 28,245

(d) Goat 1, 83,826

2. Assam

Area 78,438 sq.km

Population 2, 22, 94,562

Livestock Population

Cattle 77, 85,717

Buffalo 6, 51,819

Sheep 77,404

Goat 26, 44,704

3. Manipur

Area 22,327 sq.km

Population 18, 62,714

Livestock Population

Cattle 5, 08,262

Buffalo 95,249

Sheep 7,533

Goat 32,942

4. Meghalaya

Area 22,429 sq.km

Population 17, 60,626

Livestock Population

Cow 1, 89,000 (Non-descript)

Cow 10,000 (Cross breed)

Buffalo 10,000

Total milk production – 59,000 tonnes up to 1997-98

Milk Plant	6 Nos.
Feed unit	2 Nos.
Breeding farm	4 Nos.

5. Mizoram

Area	21,081 sq.km
Population	6, 86,217

Livestock population

Cattle (Male)	12,025
Cattle (Female)	40,856
Buffalo (Male)	2,102
Buffalo (Female)	5,345

6. Nagaland

Area	16,579 sq.km
Population	12, 15,573

Livestock population

Cow	70,000 (Cross breed)
Cow	9,000 (Non-descript)
Buffalo	9,000

7. Tripura

Area	10,486 sq.km
Population	27, 44,827

Livestock population

Cow	2, 55,000 (Non-descript)
Cow	45,000 (Cross breed)
Buffalo	7,000

References:

All India Dairy Business Directory (Dairy Year Book), 2001, Special Millennium Issue, Sadana Publishers & Distribu.

Meghalaya Achievements in 30 years 1972-73 to 2001-2002; Govt. of Meghalaya, Directorate of Economics & Statistics, Shillong, 2002.

Meghalaya Socio-Economic Review : Government of Meghalaya, Directorate of Economics and Statistics, Shillong, 2003.

Mizoram Statistical Hand Book 2000, Directorate of Economics & Statistics, Govt. of Mizoram, 2001.

Mudgal, V.D., Singhal, K.K. and Sharma, D.D. Advances of Dairy Animal Production, International Book Distributing Co., 2003.

Pocket Statistical Hand Book Meghalaya, Directorate of Economics & Statistics, Govt. of Meghalaya, Shillong, 2003.

Ramasamy, D., Shibu, A.V. and Gopi, H. Dairy Technologist's Hand Book, International Book Distributing Co., 1999.

Statistical Hand Book of Manipur, Directorate of Economics & Statistics, Govt. of Manipur, Imphal, 2002.

Statistical Hand Book of Nagaland, 2000, Directorate of Economics & Statistics, Govt. of Nagaland, Kohima, 2001.

Venkatasubramanian, V., Singh, A.K., Rao, S.V.N. Dairy Development in India, Concept Publishing Co., New Delhi, 2003.

Where do we Stand in 2003 Meghalaya and North East and India and the World: Govt. of Meghalaya, Directorate of Economics & Statistics, Meghalaya, Shillong, 2004.

Winton, A.L. and Winton, K.B. Milk and Milk Products, Agrobios, India, 2000.

* * *

10

FINANCING ISSUES IN AGRICULTURAL SECTOR: AN IMPLICATION

JAYNAL UD-DIN AHMED
Associate Professor, Department of Management,
North Eastern Hill University, Tura Campus, Chandmari,
Tura - 794002, West Garo Hills, Meghalaya, India,
E-mail: jaynaluahmed@yahoo.co.in / juahmednehu@gmail.com

ABSTRACT

It is observed that despite mounting overdues in each year, the agriculture sectors enjoyed an incredible share of bank credit. However, the undeniable fact remains that recovery performance in agriculture and allied activities is relatively discouraging to other sectors in the district. The unsatisfactory performance of agricultural credit indicates a wide gap between deployment of bank credit and agricultural development in the area under study. Thus the paper has also identified some of the issues like C/D ratio, mounting overdue, business volume (deposit plus advances), and interest rate on agricultural credit etc that affect the deployment of bank credit in the area under consideration. It may be suggested that there should be proper recovery of loan amount from loanee in order to revamp the performance of credit delivery institutions in the area under study. This would lead to nourishment of a new and active agricultural sector in the district with positive strokes.

Introduction:

The finance implies the funds borrowed and is thus the counterpart of debt (Basu, 1995). It is an input and acts as the accelerator of development. Bank finance or bank credit may be defined as the activity concerned with the raising and administering funds from banks. Agriculture in India is central to growth strategy for socio-economic development. A sustained increase in agricultural production and productivity is vital for achieving the socio-economic objectives. In order to raise the agricultural production, Sadhu and Singh (1995) stressed the need for availability of credit to farmers at appropriate costs. Agricultural credit thus is a nucleus of the system of farm operation. In an agrarian economy like India where 79 per cent of the workforce and 64 per cent of the geographical area are under agriculture, it is necessary that desired quantum of credit should be made available in rural areas. Sufficient and timely availability of credit particularly to the small farmers is of crucial importance for a dynamic agricultural sector. However, the poor farmers often lack the accessibility to institutional sources of finance and have to obtain loans from unscrupulous money-lenders at high interest rates (Ahmad, 1998). It was reported that in the process of economic development, credit was provided at concessional rates of interest mostly to big farmers; loans were wrongly and unprofitably allocated or misused; standard credit packages advocated were not suited to the heterogeneous needs of small farmers (Shete, 2002). The reform measures launched in 1991 based on Narasimham Committee (1998) indicated that timely and adequate availability of credit rather than its costs is very indispensable for intended beneficiaries. The impact of Narasimham committee recommendations with regard to lending to priority sector in general and agriculture and allied sector in particular has not been encouraging (Majumdar,2001). It appears that there is a growing anxiety that the process of financial sector reforms has by-passed the agricultural sector. In this respect Y. V. Reddy (February 3, 2001), Deputy Governor of RBI, remarked that the flow of credit to agriculture/ rural areas has not been up to the mark owing to accumulation of losses in public sector banks on account of mounting non-performing assets(NPAs). The banks are not able to reach the prescribed target of lending to agriculture and weaker section.

The small farmers and marginal farmers are continued to be both credit and demand constraints. The bankers have the excuse that the constraints facing by them with regard to deployment of credit in agricultural sector are lack of viable credit products, implying lack of demand for credit etc. On the other hand, there exists an informal sector which provides agricultural credit at higher rates of interest, which indicates that there are no demand constraints. Thus, it can be observed that the demand for funds for a dynamic agricultural sector is enormous. An attempt has been made in the paper to diagnose empirically the bank financing issues of agriculture and allied sector in a less developed district viz., Karimganj district of Assam.

Background of the Study Area:

Karimganj is the southernmost district of Assam covers an area of 1809 sq. km comprising 933 villages and 4 towns(GoA: 2007). The topography of the district is quadrilateral shape composed of high hills, low lands, level plains dotted with isolated hills called 'tillas'. The district is densely populated, industrially backward with an agrarian bias. The literacy rate in the district is 57.4 per cent which is higher than the national average (GoI, 2002). Nearly 72 per cent of the workforce and 50 per cent of the geographical area are under agriculture .The soil of the district is highly acidic in nature and is of alluvial origin. High lands within the district are generally planted with tea while plain areas are covered with rice. The rice and tea together occupy more than 90 per cent of gross cropped area of the district. Sugarcane, oil seeds, potatoes, pulses etc are also produced. The area under high yielding variety is 39 per cent of the gross cropped area of the district. The average yield of rice per hectare was 1,542 kg as against 1,784 kg in the national level during 2003-04. The climatic condition of the district is characterized by high temperature and heavy rainfall. The average minimum temperature recorded 31.6^0 centigrade and minimum 16.1^0 centigrade; rainfall varies from 19.6 mm to561.1 mm during 2000-01. The adequacy of finance is essential for development of a dynamic agricultural sector. The rate of growth of flow of institutional finance in the district has not shown significant rise. The district has 4.3 per cent of bank offices of state Assam (GoI, 2005). At present, 9 public sector banks with

32 branches, 1 regional rural bank (Cachar Gramin Bank) with 17 branches and 2 co-operative banks with 3 branches are operating in the district. The table- 1 shows the banking network of the area under study.

The economy of the district is basically agrarian. The economic development of the district is highly dependent on the development of agriculture and allied sector, agro-based industries and service sector. The agriculture in the district is mainly subsistence type which has hampered the industrial growth.

Table 1: Banking Network of Karimganj District as on March 2007

Sl.No.	Types of Banks	Number of Banks	Nos. of Branches
1.	**Public Sector Banks:**		
(a)	State Bank of India.	1	6
(b)	Other Nationalized Banks.	8	26
2.		**1**	**17**
(a)	**Regional Rural Banks:**		
3.	Assam Gramin Vikash Banks.	**2**	**3**
	Co-operative Banks.		
	Total	**12**	**52**

Source: Lead Bank Office, Karimganj District, Assam.

Deployment of Bank Credit to Agricultural Sector in India:

The commercial banks played a significant role among the institutional sources of credit for priority sector in India. The significance of bank credit in the priority sector can be observed against the backdrop of increasing quantum of such credit to this sector. The deceleration in bank credit growth is reflected the fact of slowdown in real economy as also cautious approach adopted by banks against the backdrop of growing uncertainties. The data suggested that growth rate of bank's lending to small enterprises, and agriculture and allied activities increased substantially during 2000-09. The table -2 presents total bank credit and the deployment of bank credit to priority sector in India during 2000-2009.

Table 2: Deployment of Bank Credit to Agricultural Sector in India

(Amount Rs. in crores)

Sl No	Priority Sectors	March 2000	March 2001	March 2002	March 2003	March 2004	March 2005	March 2006	March 2007	March 2008	March 2009
1	Total bank credit	4,00,818 (100)	4,69,153 (100)	5,36,727 (100)	6,69,534 (100)	7,64,383 (100)	10,40,909 (100)	14,45,837 (100)	18,48,187 (100)	22,47,289 (100)	26,48,501 (100)
2	Agriculture & allied	44381 (11.1)	51922 (11.0)	60761 (11.3)	73,518 (10.9)	90,541 (11.8)	1,25,250 (12.0)	1,72,279 (11.9)	2,30,377 (12.8)	2,75, 342 (12.5)	3,38,656 (13.0)
3	Small scale industries	52814 (13.2)	56002 (11.9)	57199 (10.6)	60,394 (9.0)	65,855 (8.6)	74,588 (7.2)	90,239 (6.2)	1,23,987 (8.7)	2,04,892 (9.3)	2,59,998 (10.0)
4	Other priority sector	34632 (8.6)	46490 (9.9)	57299 (10.7)	77,697 (11.6)	1,07,438 (14.0)	1,81,638 (17.4)	2,47,379 (17.1)	1,60,343 (8.9)	1,91,878 (8.7)	2,03,154 (7.8)

Figures in the parentheses indicate percentages to total bank credit deployed.
Source: *Reserve Bank of India, Report on Currency and Finance, Various issues*

The outstanding bank credit of small scale sector increased from Rs. 52,814 crores in March 2000 to Rs. 2, 59,998 crores in March 2009 recording around 5 fold increases but the percentage share to total bank credit has declined from 13.2 per cent to 10.0 per cent. During the same period, the amount of bank advances to agriculture sector increased from Rs. 44,381 crores to Rs.3, 38,656 crores. There is a sudden jump of bank credit deployed to other priority sector during the period under consideration. The same has recorded an increase from Rs. 34,632 crores in March 2000 to Rs.2, 03,154 crores in March 2009. It can be argued that the growth of advances to other priority sector in absolute figure has gone over the growth of bank credit to small sector although the percentage share has remained lower throughout the period. This signifies that the banking system is doing some appropriate and practical strategies in assisting to those sectors of the economy which has not received adequate support of institutional finance.

Financing of Agriculture: An Implication

The performance of banks in terms of branch expansions, deposit mobilization, credit channelization, credit-deposit ratios, population served per branch etc is essential for enhancing the bank's lending capacity. The following table-3 has reviewed the operational activities of banks in the area under study.

Table 3: Performance of Banks in the District

Year	Bank Branch (in number)	Deposit (Rs. In lakhs)	Deposit per office (Rs. in lakhs)	Advance (Rs. in lakhs)	Advances per office (Rs. in lakhs)	C/D ratio (in pc)	% share of priority sector advances to total advances	Population per bank branch (in thousands)
1997	50	12160	405.3	2689	89.6	21.3	63.4	34976
1998	51	15632	521.1	3475	115.8	22.0	50.8	45097
1999	51	18158	605.3	2735	91.2	19.8	58.6	37800
2000	51	24516	817.2	3287	109.5	13.4	60.9	41743
2001	51	29756	987.2	3891	133.9	13.0	59.5	43780
2002	52	31972	1032.8	4054	187.0	12.6	61.3	49741
2003	52	37531	1267.9	4328	198.0	11.6	57.3	53490

Source: File records, Lead Bank Office, Karimganj District, Assam.

It is clear from the table that the credit off take from banks in the area under study continues to remain slow and there has been a trend to attract more deposit. The analysis of bank deposits revealed that the district has witnessed a significant growth in deposit as well as deposit per bank branch. This shows a close interaction of bank personnel and rural population. This further may be due to the fact that the number of population served by per bank branches in the study area is much more than the national average. However, the flow of credit has not been commensurate with the quantum of deposits. During 1997-2003, the growth of deposits (3 fold) recorded a higher growth than that of growth of credit (1.4 fold). It has also been observed that the average C/D ratio of banks in the district is extremely lower than the stipulated ratio of 60 per cent throughout the period which indicates siphoning of loanable funds to other places outside the district.

The banks operating in the study area, keeping in view the aim and objectives for the economic development under the lead bank schemes have made efforts in providing financial support to agriculture and small scale industries. The sector wise credit deployment in the district is shown in table-4. It is observed from the table that the total loan sanctioned by the banks increased from Rs. 68.6 lakhs in March 1996 to Rs.99.1 lakhs in March 2005. The loan sanctioned to agricultural sector was Rs.24.3 lakhs in March 1996 which reduced drastically to Rs.15.4 lakhs in March 1999. The same, however, increased to Rs. 24.9 lakhs in March 2005. A wide fluctuation in sanctioning the bank loan to small scale sector has been practiced during the period under consideration. But the service sector recorded an increase of Rs.27.5 lakhs loan amount during the same period.

Table 4: Sectorwise Credit Deployment as on March

(Amount in Rs . 000)

Year	Agriculture & allied sector	Small scale sector	Trade& services	Total
1996	24,349	23,541	20,800	68,690
1997	26,965	15,833	27,754	70,552
1998	26565	16,954	34,393	77912
1999	15,419	13,771	30,339	59,529
2000	15,977	10,571	45,025	71,573
2001	17,998	19,389	41,316	78,703
2002	18,674	10,867	42,534	72,075
2003	22,397	17,650	36,867	76,914
2004	25,345	18,945	40,976	85,266
2005	24,985	25,742	48,325	99,052

Source: Lead Bank Office, Karimganj, Assam.

The performance of banks to fulfill the target of bank credit in agriculture and allied sector vis-a-vis other sectors in the area under study are presented in table -5. It is evident from the table that the banks have failed to achieve the target fixed in different sectors and there is low degree of positive correlation between target fixed and achievements. The overall position in case of activities allied to agriculture does not present an encouraging picture. The percentage of target achieved decreased from 91 percent in 1996 to 87 percent in 2003 which further increased to 90 percent in 2005.There is wide variations in the credit performance of banks in agricultural sector. However, the small scale sector and service sector have received progressively better treatment in respect of fulfillment of target by the banks. It has been observed that in case of service sector the achievements surpassed the targets defeating the objective of giving weight age to other sectors. Thus, a check is required in this respect in the reform era otherwise agriculture and allied activities would suffer from lack of funds.

Table 5: Sector-wise Credit Performance of Banks

(in percentage of target achieved)

As on December	Agriculture & Allied	Small scale sector	Trade& services
1996	91	97	143
1999	86	95	100
2001	89	99	106
2003	87	92	112
2005	90	98	108

Source: File records, Lead Bank Office, Karimganj District, Assam.

The liquidity of banks largely depends on the recovery of advances. The lower recovery indicates erosion of banks profitability and blocking up bank credit meant from developmental project of the area (Vargheese, 1994). It consequently restricts the banks lending capacity and stands in the way of dilution of funds to farm activities and hence the socio economic development of the area. Table-6 presents the recovery position of banks in agriculture vis-a-vis other sectors in the district under study.

Table 6: Sector-wise Recovery Positions of Banks

(Rupees in thousands)

Year End June	Agriculture & allied sector			Small scale sector			Trade & services		
	Total Demand	Total Re-covery	Over-dues	Total Demand	Total Recovery	Over-dues	Total Demand	Total Recovery	Over-dues
1997	42,345	6,534 (15.4)	35,811 (84.6)	12,875	3,142 (21.9)	9,733 (78.1)	1,28,765	45,468 (35.3)	83297 (64.7)
1999	45,780	7,986 (17.4)	37,794 (82.6)	14,310	3,609 (25.3)	10,701 (74.7)	1,98,230	75.586 (38.2)	122644 (61.8)
2001	49,867	8,790 (17.6)	41,077 (82.4)	23,543	6,124 (26.0)	17,419 (74.0)	2,40,856	1,06,762 (44.3)	134094 (55.7)
2003	43,876	10,145 (23.1)	33,731 (77.9)	26,839	7,601 (28.3)	19,238 (71.7)	2,35,754	96,763 (41.1)	138991 (58.9)
2005	58,735	15,945 (27.1)	42,790 (72.9)	31,427	10,295 (32.7)	21,132 (67.3)	2,65,960	1,54,390 (58.1)	111570 (41.9)

Figures in the parentheses indicate percentages to total
Source: Lead Bank Statements, Lead Bank Office, Karimganj District.

It has been observed that the recovery performance of banks was poor in the district. It was 15.4 per cent in agriculture and allied activities followed by 21.9 per cent in small sector and 35.3 per cent in service sector at the end of June 1997. However, the overall recovery of bank advances has been improved steadily during 1997-2005. At the end of June 2005, agricultural sector accredited to 27.1 per cent recovery, small scale sector 32.7 per cent and services sector 58.1 per cent leaving thereby a major chunk of bank advances as overdue.

The recovery performance in agriculture and allied activities is relatively discouraging to other sectors in the district. The severity of the situation would be cleared when we unearth Rs15, 945 thousands (27.1 per cent) was realized against the total outstanding dues of Rs.58, 735 thousands under this sector in June 2005. A noteworthy feature of banks' lending is that, despite mounting overdues in each year, the agriculture and allied sectors enjoyed an incredible share of bank credit. The reasons of poor recovery may be attributed to various factors such as lack of supervision of end use of fund , insufficient and under financing, deficiencies in lending policies, defective processing of loan applications, political interference for loan exemption, willful default, communication gap between bankers and customers, constant failure of monsoon, natural calamities famine flood etc. The mounting overdues restricted the bank's lending capacity. Therefore, an immediate action from appropriate authority to pick up the pace of recovery of loan in agricultural sector is needed.

Conclusion:

The above analysis, therefore, shows that the bank credit support to agricultural sector is not satisfactory. Thus, it is very important to identify the issues leading to low level of financing of agricultural activities. In this context the major issues viz, percentage of overdue in agricultural sector, interest rate on farm credit, performance of banks measured with Credit-Deposit (C/D) ratios, credit delivery centers, ie, branch expansion of banks and volume of business (deposit plus advance) associated with this sector has been felt. Higher recovery ensures bank to endorse higher quantum of credit to agriculture and vice-versa. The stumpy performance intensity of credit delivery institutions, ie, unhealthy C/D ratio, does present an impression that flow of credit to agricultural sector has come down substantially despite various measures undertaken. The adequacy of credit delivery centers i.e., branch expansion of banks, is also a credit supporting variable to farm activities. It may be suggested that there should be proper recovery of loan amount from loanee in order to revamp the performance of credit delivery institutions in the area under study other wise banks would suffer causing harm to this important sector of the economy as well as the economy as a whole. This would lead to nourishment of a new and dynamic farming culture in the district with affirmative fondles.

References:

Ahmed, R, **1998**. *Rural Banking and Economic Development*, Mittal Publication. New Delhi.

Basu, C.R. **1995.** *Same Facts of Bank Credit Planning and Central Banking Policy: The Indian Experience.* Arhiant Publishing Home. Jaipur.

Govt of India. **2002**. *Basic Statistics of North Eastern Region.* North Eastern Council, Ministry of Home Affairs, Shillong, Meghalaya, India.

Govt of India, **1998.** Report of the Committee on Financial System Ministry of Finance (Narasimham Committee-II). April.

Govt of India. **1991a** Report of the Committee on Financial System, Ministry of Finance (Narasimham Committee) December.

Govt. of Assam 2006 *Statistical Hand book,* Directorate of Economics and Statistics, Guwahati, Assam.

Govt. of Assam. **2005**. Directorate of Economics & Statistics, Guwahati, Assam.

Lead Bank Statements, Lead Bank Office of the district

Majumder, N.A. **2001**. *The New Architecture of Rural Credit System*, Prof. M.L. Dantwala Monograph Series, Department of Economics, University of Mumbai, August 2001.

Office of the Joint Director of Agriculture, South Assam Zone, Silchar

Reddy, Y. V., **2001** Address at the Conference of Indian Society of Agricultural Marketing at Vizag (A.P.) February, 2001.

Reserve Bank of India, Report on Currency and Finance, Various issues

Sadhu, A.N. and Singh, A., **1995.** *Fundamentals of Agricultural Economics,* Himalaya Publishing House, Bombay.

Shete, N.B. **2002**. *Priority Sector Advances By Public Sector Banks during the post Reform Period (1992-93 to 2000-01).* Working Paper, NIBM, Pune, March 2002.

Vargheese, R. **1994.** *The Menace of Overdues in Lending,* in Nair, C.N.P (ed.) Recent Trends in Banking. DPH Publishers, New Delhi.

* * *

11

SUPPLY CHAIN MANAGEMENT IN THE INDIAN POULTRY MEAT INDUSTRY

K.Hriiyio Kayina
PhD Scholar, Department of Management,
North-Eastern Hill University, Tura Campus-794002, Meghalaya

ABSTRACT

The world poultry meat industry has been the fastest growing among different activities in the livestock sector. Consumers are increasingly preferring poultry meat to bovine and ovine meat. This increased preference may be attributed to growing concerns related to health, hygiene, quality, safety, availability, convenience, and concerns over red-meat borne illnesses. Indian consumers are no different, chicken is the major meat consumed in India and has wider regional acceptance. The main drivers for poultry meat industry, are that it is cheaper among other meat; does not have any religious association (unlike beef & pork); is available throughout India in any season and is seen as a rich source of protein with low cholesterol. Although the production and consumption of poultry meat is growing at a rapid rate, transforming gradually from unorganized, small-scale backyard farming into large scale, organized farming, there exists a number of problems such as high costs, unhygienic production of meat, contamination, poor infrastructure, unscientific processing, absence of cold chains, poor packaging and near absence of meat safety management systems. The single most important constraint however, is the highly inefficient supply chain. The purpose of the present paper is to highlight that a supply-chain based perspective can help in addressing questions of operational efficiency and effectiveness such as enhanced level of delivery performance, elimination of waste, reduction in costs, enhanced level of capacity utilization and higher returns on investment. Besides, consumers will be benefitted in terms of lower prices, higher varieties, better qualities and convenience.

Supply Chain Management:

Today's meat industry relies upon a web of inter-organizations relationships. There has already been a practice, in the business and management world, of integrating organizations to all levels of suppliers and customers.

The supply chain is defined as the network of interconnected and interdependent organizations that work mutually and co-operatively to manage, control and improve the flow of materials and information from suppliers to end users (Christopher, 1998). In agribusiness we speak of a 'Supply Chain' when actors are linked from 'farm to fork' to achieve a more effective and consumer oriented flow of products. Such supply chains may include growers, pickers, packers, processors, storage and transport facilitators, marketers, exporters, importers, distributors, wholesalers and retailers. Agri supply chains are also economic systems which distribute benefits and which apportion risks among participants. Thus, supply chains enforce internal mechanisms and develop chain wide incentives for assuring the timely performance of production and delivery commitments (Iyer and Bergen, 1997; Lambert and Cooper, 2000).

Supply Chain Management (SCM) is a management process that comprises the stages involved in fulfilling a customer order (Chopra, 2006) – i.e., integrated and synchronized operational efforts to improve the flow of material and product. Ayers (2001) has defined SCM as a value chain that involves flow of physical, financial, information/knowledge resources and spans procurement, manufacturing and distribution. SCM is an approach for organizations to control and integrate the business functions related to the manufacturing of products, including procurement and distribution, with the dual objectives of attaining higher levels of customer satisfaction and reducing inventory risk (Das and Abdel-Malek, 2003). Managing supply chains requires an integral approach in which chain partners jointly plan and control the flow of goods, information, technology and capital from the suppliers to the final consumers and vice versa. In order to react effectively and quick to consumer's demand, SCM is consumer oriented. It aims at coordination of processes (Lambert and Cooper, 2000).

SCM results in lower transaction costs and increased margins. Because of the many activities and aspects involved it demands a multidisciplinary approach and sustainable trade relations. Supply chain partnerships are based on interdependence, trust, open communication and mutual benefits. SCM helps organizations achieve a number of objectives such as serving the right customers, identifying the right suppliers and building trust with partners in the supply chain. SCM has gained much importance of late as it helps in reducing costs, increasing sales, better information flow, better control of product safety and quality, improves market positions and thus become a source of competitive advantage.

Indian Poultry Industry:

Indian Poultry Industry is one of the fastest growing segments of the agricultural sector today in India. While the production of agricultural crops has been rising at a rate of 1.5 to 2% per annum, the production of eggs and broilers has been rising at a rate of 8 to 10% per annum. (*Source: http://www.indianmirror.com/indian-industries/poultry.html*). India is the world's fifth largest producer of eggs and ninth largest producer of poultry meat. The organized sector of poultry industry contributes 70% of the total output and the rest 30% in the unorganized sector (*Source: http://www.gktoday.in/2009/12/indian-economy-meat-poultry-industry-of.html*). The poultry business in India now employs around 3 million people, some directly and others engaged in marketing and sales of poultry products. India has emerged as the only country in the developing world a self-reliant, technology driven industry, with capability to produce every essential input for successful poultry farming including indigenous genetic resource and breeding, world class poultry vaccines and medicines, specific pathogen free eggs (SPF), farms and hatchery automation systems, egg processing, poultry processing, nationwide network of disease diagnostic laboratories and facilities for entrepreneurial development and training in both private and public sectors.

The processed meat industry in India is growing at a very slow pace. Two main reasons for this are consumers' preference for live chicken and scepticism about processed chicken, and inadequate infrastructure like lack of cold chains etc. To Indian consumers, chicken is fresh if it is live and cut before their eyes, even in a very unhygienic manner. Since live birds are available in plenty in the markets, consumers prefer live birds over processed chicken, which they perceive to be "not as fresh". Also, as processed chicken is costlier than live chicken, buying of processed chicken has so far been confined to the upper income group. The change in consumer mindset in favour of processed chicken would gradually evolve with promotion and awareness. The absence of cold chains from the processors to the retailers and the lack of adequate refrigeration facilities at the retail outlets result in deterioration of the quality of processed meat, ultimately affecting consumer health.

Although, the Indian Poultry industry has made rapid progress during the last four decades a major section of the industry still suffers from a number of problems such as lack of adequate infrastructure like storages, transport, cold storage chains; unorganized and inefficient marketing system; interference of many market intermediaries; unavailability of proper nutritious feed in absence of corn or maize which constitutes the bulk of broiler feed; unawareness of hygienic practices.

The market environment presents a number of opportunities as well as challenges for the Indian meat industry. With increasing urbanisation, changes in lifestyles, and a shift in eating habits (with non-vegetarianism becoming popular), India's meat and meat products market is expected to experience growth. Rising consumer incomes of the middle class Indian, presence of private players, rising demand of the Indian poultry produce in the export market are some of the contributing factors to the growth of the industry.

SCM in the Poultry Meat Industry:

Poultry meat is a perishable product; therefore it cannot be stocked as inventory like other non-perishable products. Neither can it be made to order as the lead time for production of poultry meat (from hatching eggs to rearing fully grown broilers) is about 10-12 weeks. It is further affected by the need for immediate shipment, regardless of price. Another problem is the uncertainties in demand that is seasonal in nature and the preference of consumers for specific body parts. An additional complication is the maintenance of quality through the chain. Some products deteriorate in transit and then again in the hands of the retailer. Hence there is a need to create a lean, agile and flexible supply chain that not only leverages scale efficiency, but also meets delivery requirements effectively. The supply chain perspective provides a market oriented strategy which can be adaptive and efficient.

The scope of the field of marketing can be examined from different angles, producers'/farmers' interest, consumers' interest and traders' interest. Farmers use their available scare resources with a look out for the most rewarding benefits, they want to produce at a low cost and get good prices for their product. Consumers will look to satisfy their needs in the most convenient ways as in low prices, prompt delivery, quality and safety. Traders through the process of marketing not only fulfil the needs of the producers and consumers, but also make out their living. These conflicting interests often result inefficiencies and ineffectiveness in the marketing system. SCM and the concepts like information sharing, data integration and consolidation, flow of funds, matching demand with supply, collaborative planning, forecasting and replenishment can be of effective use in the poultry meat market. Application of SCM will thus allow assigning of activities to the level that best enhances performance areas such as communication, coordination and service.

The supply chain of a poultry meat industry includes entities such as poultry farms, meat processing units, distributors, retailers and end consumers. A typical poultry meat supply chain may be as depicted below,

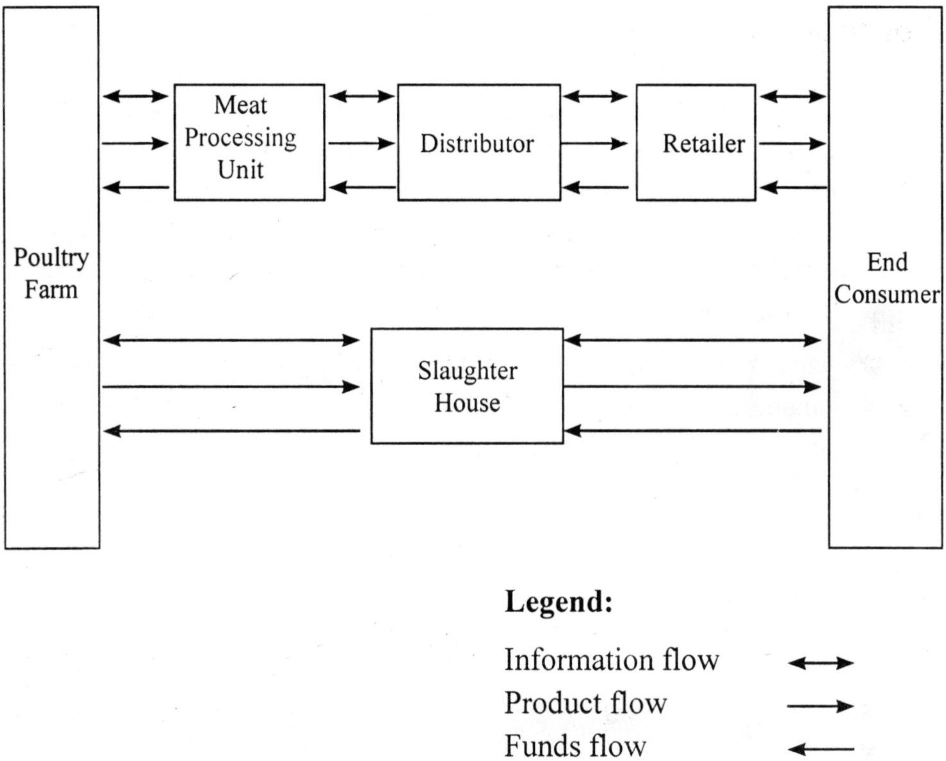

The integration of these entities and the relationships therein will determine how activities are carried out for managing the poultry meat supply chain efficiently and effectively. Some of the importance features are, Supplier partnership - deals with the long term relationship between the focal organization and its suppliers. These partnerships yield increased market share, reduce lead time, improve service delivery and quality, shorter product life cycles of meat products, free exchange of information and coordinated decision making (Corbett et.al. 1999), Customer Relationship – involves the practices of managing downstream supply chain partners and maintaining profitable relationships, information sharing – refers to sharing of critical and proprietary information communicated among supply chain partners. Proper management of these factors will provide for competitive advantage in terms of cost, quality, delivery, and thus increase market performance.

Conclusions:

The poultry meat industry is an increasingly competitive and complex market. It is continually requiring technical knowledge and management competences to attend to consumers growing concerns. These have created a need for producers to considerably improve their products and increase competitiveness of their chains. Although production and consumption of poultry meat in India are growing at a very rapid rate, the industry is still characterized by inefficiencies. Therefore the need of the hour is look at it from a managerial point of view for better control and better coordination in the poultry meat supply chains.

References:

Ayers, J.B. (2001), Handbook of Supply Chain Management. CRC Press, Boca Raton.

Chopra, S. and Meindl, P. (2006), Supply Chain Management, Pearson Education, New Delhi.

Christopher, M. (1998), Logistics and Supply Chain Management: Strategies for reducing cost and Improving Service. Financial Times, Pitman Publishing, London. P.19.

Corbett, C.J. et.al. (1999) Partnerships to improve supply chains, Sloan Management Review, Vol.40, No.4, pp.71-82.

Das, S. and Abdel-Malek, L. (2003), Modelling the flexibilities of model quantities and lead times in Supply Chains, International Journal of Production Economics, Vol.85, No.2, pp.171-181.

Lambert, D.M. and Cooper, M.C. (2000), Issues in Supply Chain Management. Industrial Marketing Management vol. 29, pp. 65-83.

Mitra, S. and Bose, D. (2008), Supply chain issues in the Indian poultry-meat industry: the case of a vertically-integrated farm. Working Paper Series, N.575. IIM Calcutta.

Pereira, S.C.F. and Csillag J.M. (2003), Supply Chain Performance Measurement Systems: A study of meat supply chain in Brazil. Production and Operations Management Society, Atlanta.

Prakash, G. (2008), Supply Chain Management in the Indian Meat Industry. Productivity, Vol.48, No.4, pp.415-423.

Schotzko, R.T. and Hinson R.A. (2000) Supply Chain Management in Perishables: A Produce Application . Journal of Food Distribution Research. Vol.31, pp.17-25.

* * *

12

WORLD TRADE ORGANIZATION AND INDIAN DAIRY INDUSTRY

Dr Ashish.K.Makwana
Assistant Professor, Dairy Business Management Department College of Dairy Science Anand Agricultural UniversityAnand-388110,Gujarat.
E-mail: ashish_gau@rediffmail.com

ABSTRACT

The positive role that dairying could play in providing income and employment opportunity was clear to policy-makers long time back and a set of measures were put in place to develop and protect the dairy industry. Immediately after India gained independence, the Milk Control Board was set up which controlled the supply and distribution chains. This however led to emergence of a set of middlemen and the share of producers in the sales declined. With processing units set up in cities, it became difficult to procure and transport milk from the centers of production which were in the rural areas. As a result, the yield of milk declined and imports of milk powder went up. While the government was grappling with these problems, a quiet revolution was taking place in the village of Kaira in Gujarat which set up a cooperative to procure process and market milk. In 1973, the Kaira Cooperative Union set up a marketing agency named Gujarat Cooperative Milk Marketing Federation (GCMMF), which follows a three-tier structure at village, district and state level for procuring, processing and marketing milk and milk products. The district units also provide technical back-stopping to the milk producers and a range of services such as feed, veterinary care, artificial insemination, education and training. These milk cooperatives of Gujarat today own the GCMMF, the largest food products business in India. GCMMF is also the largest exporter of dairy products from India and its brand name Amul is known all over the world. This experiment laid the foundation of the cooperative movement in milk production and marketing in India. The federal and egalitarian structure of these cooperatives ensures social and economic equity to the milk producers and is one of the major reasons for its success. The government adopted this successful model and set up the National Dairy Development Board (NDDB) in 1965 which prepared a blueprint for a milk revolution across the country.

Introduction:

It has been eventful journey for dairy industry of India from the era of scarcity at the time of independence to largest milk producing country in the world. India's milk production in 2009 was 104.8 million tonnes and per capita availability of milk was 252 gm/day. Indian dairy industry has helped both producers and consumers. Dairying has brought socio economic transformation of rural India. The dairy sector is characterised for its complimentary with agriculture and act as a source of protein rich diet for vegetarians. Due to well established cooperative network milk producers get remunerative price of milk throughout the year. Consumers also get quality milk products throughout the country at reasonable price.

Characteristics of Indian Dairying:

India has a distinct pattern of production, processing, marketing and consumption of milk, which is not comparable with any large milk producing country. It happens through organized and unorganized forms. Approximately 70 million rural households, primarily small and marginal farmers and landless labourers are involved in milk production. Under organized form farmers are organized into village Dairy Cooperative Societies (DCSs). Milk production starts as a trickle of one to two liters per household. A unique collection system transforms this trickle into a veritable flood of processed milk for urban consumers. The total milk handled by DCSs form part of national milk grid which links the milk producers with consumers throughout India bridging the gaps on account of seasonal and regional variation in the availability of milk.

Focus of world:

The world dairy industry is zooming in on India due to its fast growing markets. The changing international dairy trade pattern and the world trade organization (WTO) offer both challenges and opportunities to Indian dairy industry. In India, dairy profession is practiced in totally different way then practiced in developed western nations. In India, milk producers keep one or two milch animals where as in western countries big dairy farms are a general practice. Irrespective of these differences, Indian dairy industry is now centre of attraction from world due to its low cost economy, liberalization policy of India, cheap labour, availability of technical manpower, large population and hence huge market etc.

Trade in WTO regime:

WTO came into existence on 1st January 1995. India is a member of WTO. WTO was established to have distortion free trade among member countries for the benefit of all the member countries. WTO has devised 3 step policy reforms. Aggregate measure of support (AMS) was first to be computed in agriculture taking 1986-88 as the base year. Secondly AMS levels were to be frozen at existing level and in the third step AMS were to be reduced by 20% between 1995-2000 by the developed nations and by 13% between 1995-2004 by the developing countries.

The WTO member countries can be divided into developed, developing and underdeveloped nations. The working of WTO so far has shown that developed nations have violated many of the provisions of

WTO commitment including dairy sector also. Developed nations resort to providing subsidy to their farmers/ milk producers under one or the other pretext leading to unfair market competition, adversely impacting milk producers in India. Apart from price, developed nations also fix arbitrarily very high quality standards for milk and milk products under sanitary and phytosanitary clause. Thus it makes practically impossible to export milk and milk products from developing nation like India. In India the share of buffalo milk is higher compared to cow milk in total milk production. At international level there is also some sort of opposition to buffalo milk as propagated by developed nations in which higher proportion of cow milk is produced.

The speed of implementation of reforms is also very slow. The distorted policies have adversely hit the developing nations including India. Indian dairy industry suffered from highly subsidized imports of dairy products from EU and USA. The developing countries are thus taking tough stand at various WTO ministerial meetings.

The global recession witnessed by world has already impacted Indian Dairy industry. It has led to reduction in exports from India. The international prices of milk have gone down and it is likely that still it will go down. In such a recessionary period EU has stopped providing export subsidy. EU and USA farmers demand that their milk should be purchased at guaranteed price and supplied to poor countries as food aid.

Under domestic support, The WTO rules identify different forms of trade distorting farm subsidies which all together are grouped under overall trade distorting support (OTDS). Amber Box (AMS), de minimise and blue box. Non trade distorting farm payments are covered into green box. No limit is set for them. However large parts of farm subsidies are now shifted to green box.

India's position in global trade:

As milk is perishable in nature only seven percentage of world production is traded excluding trade between European Union Countries. The major dairy products traded in world market include butter cheese condensed milk, whey, casein, etc. The international trade in dairy handles approximately six percent of total milk production of the world. The major countries dominating the trade include Unites States of America (USA) & European Union (EU) which constitutes 50 percent, Australia and Newzeland constituting 45%. And only 05% is attributed to rest of the world. India even though largest milk producing country in the world is part of this rest of the world countries. The proportion of different dairy products traded in international markets are shown in table no.1

Table 1: International Dairy trade (Million tonnes)

Products	Intra EU Trade	World Trade	World Population	World Trade as percentage of production
Liquid milk	1.7	<0.5	>100	<0.5
Cream	0.1	0.1	-5	-2
Yogurt fermented milk	0.75	0.4-0.5	-15	-3
Cheese	1.8	1.1-1.2	15	7.8
Condensed Milk	0.2	0.5-0.6	4	10-15
Butter	0.6	0.8-0.9	6.5	10-15
Whole milk powder	0.25	-1.3	2.5	5 0
skim milk powder	0.5	-1	3.5	3 0

Source: Dairy India Year book, 2007

India is not a major player in international trade. India's share in world trade of dairy products is very less (0.2 %). The exports of dairy products from India in 2007-08 were of about 178 Million. But it is required to lay emphasis on (a) improving quality of raw milk, (b) dairy hygiene standards (c) maintain consistent export policy (d) use of improved packaging systems.

Although India's share in global dairy trade is less, it has two unique competitive advantages. (a) Low farm gate price (b) Proximity to milk deficient countries. Overall 70 percent of world's milk is more costly than India. India's geographical location is in vicinity of milk deficient countries of Asia and Africa which are imports 40 to 97 percent of their milk requirement. However due to low quality, Poor hygienic standards and lack of marketing experience, India is not able to gain advantage. Intensive efforts are needed to meet the WTO's sanitary and phyto- sanitary (SPS) and Technical Barriers to Trade (TBT) agreements. India's export potential is shown in Table 2.

Table 2: India's export potential

Existing products	New products	Emerging products
Milk powders, Traditional products	Milk fat cream, Condensed milk containing sweetening matter, concentrated liquid or semi liquid whey, Dry blocks or powdered whey, Dairy spreads, Processed cheese	Condensed milk not containing sugar or sweetening matter, other fermented products, powered cheese, natural milk constituents, Butter oil, Probiotic products

Impact of WTO on Indian dairy sector:

The experience has shown that WTO framework aims at imposing discipline on producers so that consumers get the maximum benefits. The WTO's proposition is that free trade is beneficial to all the countries. But this happens only in the situation of the perfect competitive scenario. In reality it is difficult to run all the businesses in perfect competition. International agriculture markets are thin and dominated by few large players. This kind of distortion is of serious concern to India as, the thin oligopoly international market may have a magnifying impact on world prices even with a minor change in production and consumption condition. The oligopolistic multinational companies and dairy trade are economically and politically strong and can eliminate small companies of developing world.

The relative advantages and disadvantages of trade in dairy products WTO for India are shown in Table No 3.

Table 3: Advantages and disadvantages of WTO trade

Advantages	Disadvantages
Low cost of milk production	Poor microbiological quality of milk
Well established dairy cooperative federations have greater potential for investment in other countries.	Adverse impact on environment as large scale polluting multinational corporations find loopholes in regulatory rules in developing country like India
Increased availability of capital due to difference in exchange rate among nations in international trade	Greater risk of transmission of diseases among nations
Access to new technology	Lack of skill in handling milk on commercial scale by using modern equipments
Increase in purchasing power of consumers due to economic progress	Highly vulnerable small holder milk producers

Conclusion:

Government of India and the non government dairy organizations should establish appropriate network to gather market information and intelligence, monitor international dairy market trends and take timely corrective steps to protect Indian dairy industry from unfair competition. Government of India should take firm stand at various WTO ministerial meetings in the larger interest of large number of small and marginal milk producers.

References:

Abdul Kalam, A.P.J. (2009) "Address by Chief Guest" *Indian Dairymen* 61 (3) pp12-15

Bhasin, N.R (2009) "Address" *Indian Dairymen,* 61 (3) pp 11.

Datta, S.K (2007) 'Challenges in marketing or dairy products' *Dairy India 2007*, pp67-69.

Khanna, R.S (2007) 'WTO and the Indian dairy industry' *Dairy India 2007*, pp-83-85

Sahai, S.K. (2009) " Chief guest address" *Indian Dairymen*, 60 (3) pp 29-31

Sharma, V.P (2009) " Colllapse of global trade talks: which way we are heading" *Indian Dairymen* 61 (3) pp 19-22.

* * *

13

PARTICIPATION OF WOMEN IN LIVESTOCK PRODUCTION IN INDIA

Biswajit Paul
Lecturer (Selection Grade), Department of Zoology,
Don Bosco College, Tura
E-mail: bptura@yahoo.co.in

ABSTRACT

Women have always proved to be in the helm of activities once the opportunity is meted out to them and in matters concerning Livestock it's no different. This paper brings forward the involvement and efficiency of women in the concerned field throughout the world, India in general and North-East India in particular.

Keywords: Livestock, women, food, production.

Introduction:

For poor people, there are many benefits from keeping livestock – from the largest water buffalo to the smallest insect. Livestock are a form of currency. For many people, animals represent savings. The sale of livestock and manure can mean quick cash in hard times. Income from livestock and their many products – milk, eggs, meat, wool, leather, honey – can allow poor families to put food on the table, improve their nutrition, send their children to school and purchase medicine for themselves and their animals. Livestock also act as a kind of social glue. Loans and gifts of livestock connect people to other family members, as well as to communities and institutions. In many societies, bride dowries are paid in the currency of livestock. Herders who share livestock with their relatives also share the risks brought by drought and disease. They carry heavy loads, help plough fields and provide means of transportation. Their manure fertilizes the soil. Most livestock graze on straw, grass, kitchen scraps and other waste, and thus convert unusable materials into high-quality food for humans. Their meat adds protein to cereal-based diets and can improve the nutrition of children. The presence of livestock reduces the need for human labor in the fields.

India has a large livestock population, regarded by some as an asset provided in plenty by nature, and by others as a burden. Since 1971, when 'poverty eradication' became the main theme of development planning, Indian Government recognized the livestock development program as an important tool for poverty alleviation. The focus of the programme has been on improving production of livestock commodities for income generation. In India, underprivileged families contribute 70 to 80% of the total livestock produce and livestock are central to their livelihoods and culture. Hence, to improve the livelihoods of the underprivileged families we need to understand their way of life, livestock production systems and their perceptions and about the role of livestock in their livelihood. Livestock sector includes the domesticated animals cows, buffaloes, goat, sheep, duck and poultry. It is an integral part of the farming system all over the world more specifically in the developing countries (FAO, 2007). It is an essential sector of rural economy in developing countries (Moser, 2007). It serves both the purposes of generating income (Carletto, 1998) and also provides necessary protein – one of the essential constituent of nutrition (Heffernan et al, 2001).

In all developing countries, people live side by side with livestock. Animals are reared in nearly all ecosystems, from arid high-mountain zones to low-lying deserts. People benefit from livestock in many ways:

* Livestock are used for food, income and draught power. Their skins are used for housing, clothing and household utensils and their dung is used for fuel and manure.
* Livestock often provide a substantial proportion of household wealth; they are the key asset in many dry land environments.
* For poor families with just a few chickens or goats, livestock-derived foods (such as eggs and milk) are an important source of nutrition, especially for children and mothers (Rosegrant and Thornton, 2008)

The 2008 World Development Report on Agriculture surveyed 14 countries and noted that most rural households, and 40 percent of the poorest households, own livestock. The World Bank estimates that livestock are the main livelihood asset for up to 200 million pastoralists and agro pastoralists in arid and semi-arid environments worldwide. Furthermore, 35 to 90 million of these people are extremely poor.

The growing demand for food

Population and economic growth in developing countries are increasing the demand for food, particularly meat and milk. The growth in food consumption is shifting from developed to developing countries. The IMPACT model (see Figure) projects that under a 'business-as-usual' scenario:

- Annual meat demand will increase by 6 to 23 kilograms per person worldwide by 2050
- The absolute increase will be fastest in Latin America, East and South Asia and the Pacific, with demand doubling in sub-Saharan Africa
- The demand for maize and other coarse grains for animal feed will increase global cereal demand by 553 million metric tons

between 2000 and 2050– nearly half of the total increase in demand for that period.

Figure 1: Projected per capita consumption of meats, 2000 and 2050

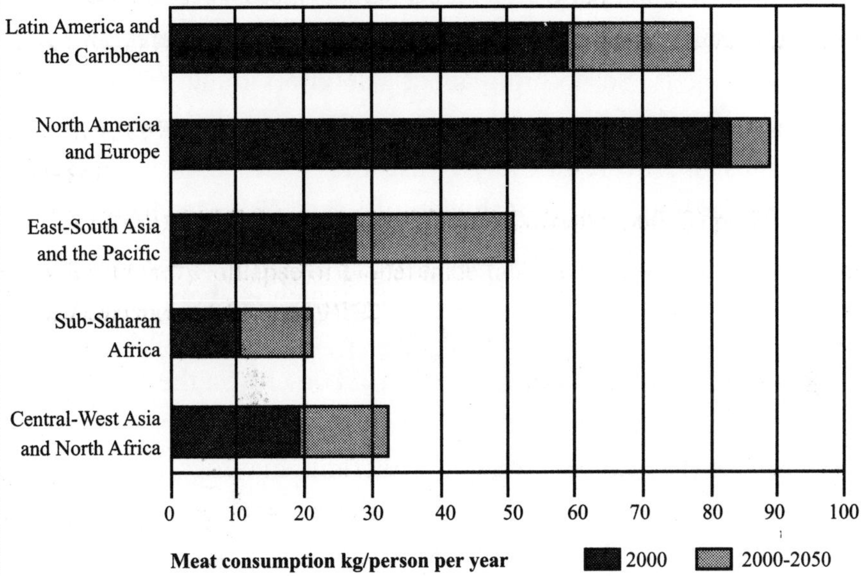

Source: International Food Policy Research Institute – International Model for Policy Analysis of Agricultural Commodities and Trade (IFPRI IMPACT) projections, September 2007

With this strong demand, the model projects that livestock populations will also increase rapidly. Between 2000 and 2050:

- The global cattle population will increase from 1.5 billion to 2.6 billion
- The global goat and sheep population will increase from 1.7 billion to 2.7 billion.

Impacts of growing food demand and supply

Considerable opportunities for livestock growth exist, but there is a danger that smallholder producers and other poor, livestock-dependent people may not be able to take advantage. This is because their access to markets and technologies is constrained. Long-term policies will be necessary to ensure that the development of livestock systems plays a role in reducing poverty, as well as mitigating negative environmental impacts, encouraging income equality and supporting progress towards reducing malnutrition. For example, policies will be needed to ensure that small-scale farmers can produce safe livestock products and sell them in appropriate markets. So, to bring about a solution to the growing demand of the Livestock sector and its impacts on the poor, the women folk have to play a substantial role in every household involved with Livestock. In India, during 2001-02, the contribution of livestock sector to the country's GDP was 5.59 percent. The value of output of livestock and fisheries sector was estimated to be around 179544 crores, which is about 27.7 percent of the total value of output of Rs.648122 crores from agriculture and allied sectors. With an estimated output of Rs. 103804 crores, the contribution of milk was higher than paddy (Rs. 73965 crores), wheat (Rs. 43816 crores) and sugarcane (Rs. 28592 crores). Milk and livestock sector provides regular employment to about 11 million in principal status and 8 million in subsidiary status. Women in animal husbandry and livestock sector constitute 69 percent of the labour force as against 35 percent in crop farming (Economic Survey 2002-03).

Traditionally, women are involved in agriculture and livestock production activities in rural areas (Waters – Bayers, 1985). Reports are there from many parts of the world that women work in livestoek in a better way then men (Ishani, 2004). According to Verma (1992), Animal Husbandry is predominantly a male affair in case of high, economic status as majority of them employ, permanent male labour to look after the animals, whereas it is predominantly a female affair in case of farmers of medium and low socio-economic status. On an average, a woman devotes 3.5 hours per day for animal husbandry activities against only 1.6 hours per day devoted by men in this category. The prosperity and growth of a nation depends on the status and development of its women, as they not only constitute nearly half of its population, but also positively influence the growth of remaining half of the population. Livestock enterprise provides employment and economic support to rural families who are landless and those possess some land. Many of the important tasks in animal husbandry activities are performed by women besides fulfilling their responsibilities as home makers (Randhawa and Chandra, 1993). Sanwan et al. (1990) reported that there is a distinct sphere of participation between males and females in farm and dairy sector, the men were the planners and women implemented the activity. Women involvement in maintaining livestock is more critical than men (Carnay, 1992). But, there always exists a division of labor in livestock management where size of the animals is a criterion. Men are more involved in maintaining large sized livestock (cow and buffalo) compared to their fairer counterparts who remain mostly with the small livestock like sheep, poultry etc. (Bravo – Baumann, 2000). Though the male persons own the large animals the women are seen to have more control over the large animals (Flintan, 2003). Men are usually involved in managing and supervision of a livestock farm and also are involved in the selling and slaughter of animals where as women are more concerned with the 'actual caring' of the farm and its inmates. She is involved in milking, cutting fodder and providing it to the animals, animal shed cleaning, looking after the diseased animals by providing medicines and nursing them. It is seen that on an average, a women involved in livestock spends 2.5 hours/ day for livestock in her schedule of her days work where she spends 6 hours /day for household work (Bekure, 1991). In Livestock production women spend 2.2 hrs/ day/household in animal care activities only (Agarwal et. al., 1999). Though male chauvinism exists in many parts of the world in livestock sector as is true in many sectors, rural female participation rate (79.4%) is higher compared to rural men (60.8%). (UNDP, 1987).

But in spite of all this devoted activities by women in the livestock sector, they are not considered as paid work as mentioned by Food and Agricultural Organization (FAO, 1998). Women has no access to technical information source related to balanced diet, health care and sanitation, care of pregnant animals and new born calf and correct procedure of milking and feeding and least access to veterinary services, true breed survival and artificial insemination techniques (Sivakumar and Trikha, 2002). Peterson (1999) concluded that though livestock management is in the domain of women they are not given proper training and so the policies and programs

Table: Mean, Sd and Rank order of the respondents according to their roles in livestock management.

	Husband (n-451				Wife (n-451)		
Livestock management	**Mean**	**SD**	**Rank order**	**Mean**	**SD**	**Rank order**	**t-test**
Cleaning of shed	1.3	0.78	3	1.3	0.75	3	0.2[NS]
Collecting and storing animal dung	0.9	0.84	6	1.2	0.79	4	-4.9*
Bathing animals	1.3	0.75	3	1.1	0.80	5	2.6*
Grazing animals	1.2	0.82	4	1.2	0.84	4	1.5[NS]
Fodder cutting	1.5	0.72	1	1.0	0.81	6	8.2*
Fodder collection for animals	1.4	0.74	2	1.4	0.73	2	1.0[NS]
Offering feed to animals	1.4	0.74	2	1.4	0.71	2	-0.3[NS]
Watering the animals	1.4	0.75	2	1.5	0.69	1	-1.0[NS]
Milking	1.2	0.81	4	1.4	0.74	2	-3.6*
Storage of milk	0.9	0.89	6	1.2	0.85	4	-5.5*
Churning and making ghee	1.0	0.91	5	1.4	0.73	2	-8.4*
Taking animals in and out of the shed	1.3	0.82	3	1.2	0.81	3	1.0[NS]
Taking care of lactating animals	1.4	0.77	2	1.3	0.78	3	1.1[NS]
Care of sick animals	1.3	0.78	3	1.3	0.81	3	1.0[NS]
Selling milk and milk products	0.5	0.79	8	0.5	0.79	9	0.4[NS]
Making of dung cakes	0.5	0.77	8	0.9	0.89	7	-6.3*
Care of newly born babies of animals	1.3	0.80	3	1.3	0.79	3	-1.2[NS]
Poultry raising	0.9	0.87	6	1.5	0.70	1	-11.1*
Egg collection	0.6	0.75	7	1.2	0.72	4	-11.3*
Selling of ghee and eggs to villagers	0.6	0.67	7	0.9	0.72	7	-6.8*

are to be reformed. Even if the programs and policies are there the male members do not allow the fairer half to attend these programs (World Bank, 2001). The involvement of women (wives) compared to men (husbands) in livestock maintenance and production is presented in the table below:

In many countries women are often denied ownership rights for large stock (cattle, camels, horses), but 'allowed' to keep small stock (sheep, goats, rabbits, poultry). Reasons include:

- Livestock ownership patterns are linked to social class, religious systems and paternalistic cultures – this means women have weaker ownership rights than men, especially in times of stress.
- The migration of men to find seasonal work makes it harder for women in their households to use land, or access credit and technical inputs for livestock production.

Evidence from around the world shows that pastoral women's rights to property and assets are being increasingly eroded, undermining their status in relation to both the household and the wider society. Not only is their burden of work becoming heavier but their labour is increasingly being controlled by men (Joekes & Pointing 1991). These changes are due to the effects of external forces, such as commercialization, sedenterization, labour migration and the internal breakdown of socio-political systems. To date, an understanding of women's role in livestock production in developing countries has been limited by cultural biases that underestimate women's contribution. Scientists and development workers have tended to concentrate on male-oriented activities (beef production, large-scale enterprises, etc.), thus neglecting those activities that women are generally engaged in, notably, milk production, the raising of small stock and poultry, meat and hide processing, etc.

Keeping the above problems faced by women in the livestock sector the attitude of the men folk should change so that there is a greater involvement and output in this sector. So the first issue that should be addressed is what are the key challenges that are to be met by the authorities in the livestock sector?

Key challenges for women:

- Understanding the basic needs of women livestock keepers, the threats they face, their roles in the livestock sector and how these roles are changing;
- Supporting women's empowerment, looking not only at how to enable women to become more market-oriented, but in particular at how to ensure that women capture the benefits of economic empowerment;
- Supporting women's access to productive resources and main assets (water, land, fuel wood, markets, knowledge), promoting their participation in small-scale dairying, and strengthening their role in decision-making processes;
- Understanding how women influence decisions and what resources they have control over;
- Supporting income-generating activities (for instance, processing and selling of livestock, forage, aromatic and medicinal plants and wildlife products) as a way to enhance women's socio-economic position in the household and empower them to take a greater role in the community. At the same time, it is important to recognize that, because of their extremely heavy workload, women have fewer opportunities to diversify or maximize their livelihoods. Time-saving opportunities, therefore, merit special attention. (IFAD, 2009)

Entrepreneurship is one of the solutions to the growing employment among rural youth. It helps to generate employment for number of people within their own social system. This is more beneficial for women in rural areas as it enable them to add to the family income while taking care of their farm, home and livestock centered tasks. Rural women possess abundant resources to take up an enterprise. She has the benefit of easy availability of farm and livestock based raw material and other resources. Hence she can effectively undertake both production and processing oriented enterprises. But to be a successful entrepreneur, women should process certain fundamental qualities beside the support of the family and government organizations. Entrepreneurial development among rural women helps to enhance their person capabilities but also decision making status in the family and society as a whole. (Sidhu and Kaur, 2005)

In many countries it has been found that it is more acceptable to work with women's groups rather than with individual women, as in Nepal (Chaney, E. 1988). This is often the only way for poor women to obtain sufficient resources (material, capital and labour) to initiate activities. The group approach with women has been tried in various projects, including those in northern Togo (sheep), the Sudan (poultry), Senegal and Kenya (sheep, goats and milk processing), Zanzibar (poultry), India and Kenya (dairy), Nepal (goat-raising), India (dairy), and Sri Lanka and Bangladesh (chickens), and has been successful in involving women in the decision-making aspects and in giving poor women an alternative to poverty (Finney 1988).

Suggestions:

1. Women must be facilitated that they may be able to establish Livestock enterprises.
2. Women must be paid their share in livestock income.
3. Role of Women contribution regarding animal production and husbandry must be acknowledged and measurement must be taken to substantiate.
4. An important gap exists between the role of women in animal husbandry and her routine work. This gap must be acknowledged through initiating different policies.
5. Women Agrarian societies may become the most important in changing her life positively.
6. There is a dire need of extensive study on the household organization of agrarian societies with a focus on women role on livestock.
7. Women may have more knowledge and interest in the subject of veterinary science and animal husbandry than men. Therefore, they must be provided employment opportunities to work for rural women guide regarding livestock.
8. Educated women of pastoral areas must be provided scholarships for the education in livestock.

9. Livestock business must be initiated as profession; in this connection livestock department must be established specifically to address their involvement. Scholarship schemes, Diploma courses and trainings must be conducted at university level.
10. Trainings must be arranged for women at community level through female trainers extensively.
11. Women and education are complementary for each others; hence, appropriate measurements must be taken to build an educated rural woman.
12. Strengthening the capacity of rural women to manage and conserve the natural resource base at the household and the community level by raising their skills in protecting and judicially managing their environment.
13. Creating and enabling environment that helps rural poor women's and communities' access to improve knowledge, skills and technologies especially in goat and sheep production systems in particular.
14. Short and intermediate term loans must be provided to women involved in livestock rearing activities on very relaxed conditions.

Conclusion:

Rural women perform variety of operations in farm and home system and have basic indigenous knowledge, skill, potential and resources which can prove helpful to establish and manage enterprises. What they need is awareness, motivation, technical skill and support from family, government and other organization. With the right assistance they can strengthen their capacities besides adding to the family income and national productivity.

References:

Agarwal, A., Narain, S. & Sen, S. 1999. The Citizen's Fifth report; *Center for Science and Environment*, New Delhi, India.

Bekure, S., de Leeuw, P.N., Grandin, B. & Neate, J.H., 1991. Masaai Herding: An Analysis of Livestock production system of Masaai Pastoralists system in Eastern Kajiado District, Kenya. *ILCA System study, No. 4. International Livestock Centre for Africa*, Addis Ababa.

Bravo – Baumann, H. 2000. Gender and Livestock: Capitalization of experiences on livestock projects and gender. *Swiss Agency for Development and Co-operation.* Bern. Indo – Swiss Natural Resources Management Program, Orissa.

Carletto, C. 1998. Household food security and the role of women: IFAD's Experience in Guatemala; *Staff Working Paper Series on garden and Household Food Security*, No. 3. Rome: IFAD.

Carnay, J.A., 1992. Peasant woman and economic transformation in Gambia; *Development and Change*, 23(2), 67 – 90.

Chaney, E. 1988. *Subsistence Projects for Rural Women.*

Economic Survey 2002-03, Economic characteristics of Indian Population, 2003, *Office of the Registrar General*, Government of India, New Delhi.

FAO, 1998. Women, Agriculture and Rural development in Near East: Findings. *An FAO study.*

FAO, 2000. National Sectoral Report on woman in Pakistan. *FAO*, Rome.

Finney, R. 1988. The role of women in the adoption of new practices. *In Proceedings of the sixth world conference on animal production,* WAAP Helsinki, 213-31.

Flintan, F., 2003. *"Engendering' Eden Vol II. Women, Gender and ICDPs in* Africa: Lessons learned and Experiences shared. *Wildlife and Development Series No. 17, International Institute of Environment and Development*, London.

Heffernan, C., Nielsen, L. & Misturelli, F. 2001. Restoring Pastoralists: A Manual. *Livestock Production Program*, Department of International Development, London.

IFAD, 2009. Rota, A. and Sperandini, S. "Gender and Livestock: tools for design" in Livestock Thematic Papers: Tools for project design. Rome: *International Fund for Agricultural Development (IFAD).*

Ishani, Z. 2004. *Scoping study on Interactions between Gender Relations and* livestock keeping in Kisumu. *Key Gender issues in Urban Food production and food security.*

Joekes, S. & J. Pointing. 1991. Women in pastoral societies in East and West Africa. *Paper No.28. IIED*, London.

Moser, C., 2007. Reducing Global Poverty: The need for asset accumulation. *Bookings Institution press*, Washington, D.C.

Peterson, P.H. 1999. The role of women in the small holder rain fed and mixed farming in India. *Women in Agriculture and Modern Communication Technology; Proceedings of a workshop*, Tunes Handboskole, Denmark, Rangneker, S. & Dolberg, F. (ed) 25 – 37.

Randhawa, A. and Chandra, S. 1993. Changing role of home science scientists in Transferring Farm Technologies to Farm Women in Agriculture. *Paper presented at national seminar on 'Women in agriculture – developmental issues' at NAARM*, Hyderabad, Dec. 28-30.

Rosegrant, M. W. & Thornton, P.T., 2008. id21 insights 72. *Environment and Production Technology Division, International Food Policy Research Institute (IFPRI)*, 2033, K Street NW, Washington, DC 20006, USA.

Sangwan, V., Munjal, S. and Punia.P.K. 1990. Participation of farm women in farm activities. *Indian J.Extn.Edu*, 26 (1&2): 112 Tamilnadu J.

Sidhu, K. & Kaur, S. 2005. Development of Entrepreneurship among Rural Women. *Department of Extension Education*, Punjab Agricultural University, Ludhiana 141001, Punjab, India.

Siva Kumar, P.S. & Trikha, R.N. 2002. Message Design for Community wall newspaper: Participatory approach. *Manage extension Research Review. Vol III (I) National Institute of Agricultural Extension Management.* Rajendranagar, Hyderabad, India.

UNDP, 1997. Human Development Report, *Oxford University Press*, New York.

Verma, S. K., 1992. Women in Agriculture: A Socio Economic Analysis 1992, *Concept Publishing Company*, A 15-16, Commercial Building, Mohan Garden, New Delhi.

Waters Bayers, A. 1985. Dairying by settled Fulani women in Central Nigeria and some implication of dairy development. *Pastoral Development Network, 20.* ODI, London.

World Bank, 2001. Livestock development, the environment, poverty and global food security. *Draft report prepared by the Animal Resources*. Thematic Group.

* * *

14

SCIENTIFIC POULTRY FARMING

Dr. L Mandal
Professor, Department of Animal Science
Ex – Dean, Faculty of Agriculture, Bidhan Chandra Krishi Viswavidyalaya
Mohanpur, Naida, E-mail: lalmohan 2006@yahoo.co.in

ABSTRACT

Poultry farmers are facing many constraints including high mortality, mainly due to Ranikhet disease and currently also to the high pathogenic influenza in many countries. Since its occurrence in Asia in late 2003.Significant improvements in this sector can be achieved through well-designed and implemented information systems that may endow poultry farmers with necessary knowledge and skill. Setting up scientific model poultry farm to aware the stakeholders in this area is a need for today's situation. Since per capita availability of egg and chicken meat in our country is much below than the developed countries therefore scientific poultry farming may represent an appropriate system for supplying the fast growing population with high quality protein and can provide additional income to resource poor small farmers and others who are engaged in this sector. This business also require less inputs and contributes significantly to food security, poverty elevation and ecologically sound management practices of natural resources.

Keywords: Scientific, poultry farming,advantages.

Introduction:

The term "Poultry" applies to rather wide variety of birds of several species and it refers to them whether they are alive or dressed (slaughtered and prepared for market). The term applies to chicken, turkey, ducks, geese, swans, guinea fowl, pigeons, peafowl, ostriches, pheasants, quail and other same birds.

Why poultry farming is beneficial?

1. It requires minimum investment to start with;
2. It gives rapid return to profit;
3. Poultry can convert feed to food protein efficiently
4. It provides a continuous source of income
5. It requires small space
6. It gives a stabilized farm income
7. Large amount of agro-industrial by-products can be used for feeding poultry
8. It gives employment opportunities
9. Poultry droppings can be used as source of organic manure

The need of animal protein for human health is known to all. The sources of animal protein food in India are mainly fish, egg and meat. Pisiculture is not possible everywhere due to lack of water reservoir particularly in North East region and more over many people are not accustomed to fish consumption. Meat may be obtained from goat, sheep and poultry. Out of these, cheapest source is the poultry because of its easy availability. Again, among poultry, chicken plays the vital role particularly for nuclear families, because one bird is sufficient to meet the requirement. India is already in deficit of protein food supply to combat with qwashiworker/Marasmus (protein deficiency diseases)-poultry farming must be encouraged in all the states including North east Region to fill up the gap between demand and supply. Moreover, most of the farmers in our Country are interested in mixed farming. The main advantage of this system is that the byproducts of crop cultivation may be well utilized for feeding to livestock and poultry and this will enable the farmer to reduce the feed cost of animals/birds.

Poultry farming can play a vital role in improving the rural economy of North east Region also. Small and marginal farmers or the poor villagers may rear birds to meet up their family need. This will not only save their money towards purchase of costly protein food like meat and eggs for their own consumption but also can give economic support to the poor village women who are not literally equipped. The most common system of rearing for these birds is the free-range system i.e. the birds are allowed to move throughout the day and keep in safe place during the night hours-also known as backyard poultry. No extra feed is given to birds they mainly thrive on seeds of weeds, grasses, insects, earthworms and left over of the family-no extra cost towards feed and housing. This is particularly practiced in most of the rural areas of all over the country. But the system is to be somewhat modified, instead of rearing low producing deshi/indigenous birds improved breeds like (White Leghorn, Rhode Island Red, New Hamshire, Australorp etc.) require to introduced. Now-a-days hybrid layers are also available which lay large number of eggs. But in that case extra feed supplementation and good mangemental procedures are to be adopted otherwise desired results may not be obtained.

Those who are interested in commercial way of poultry farming, they must choose improved breeds with high laying capacity or broilers for good quality of meat alongwith supplemented feed. The number of birds in commercial farm is generally higher so proper housing systems is also must. It may be pucca building or thatch roofed with brick wall/earthen wall according to financial ability. The house should be well protected from snakes, rates and other wild animals and should have the facility of plenty of sunshine and must be arity. Now-a-days Nationalized Banks and NABARD are providing liberal financial assistance for commercial poultry farming. So, for commercial farming improved breed, balanced feed and proper housing system are most important along with proper disease control measures.

How to minimize the cost of Poultry Farming?

70% of the expenditure of a farm is incurred towards feed cost, utilization of different Agro-Industrial by products in the ration may help in lowering feed cost. The housing cost may be minimized by constructing simple mud-house with thatch roof this will be also comfortable during summer season in reducing heat stress. Chaffed paddy straw may be spread over the floor as litter and this may be utilized afterwards as good organic manure.

Improvement in the existing system of back yard poultry:

This may be done in the following ways:

1. Supplementary feed @ 50gm/bird/day should be provided alongwith free collection of feed materials as mentioned earlier. Plenty of fresh and clean drinking water should be provided as insufficient water intake decreases egg production.
2. Regular deworming is a must, at least in 2 months interval
3. Regular vaccination against most devastating diseases like Ranikhet,

   Fowl pox and Gumboro should be done.
4. Cockerel exchange programme:- Breeding of deshi hens with improved cockerel like Rhode Island Red, /white Leghorn, New Hampshire and Australorp will definitely increase the number of eggs as well as quantity of meat.

By providing proper feeding, housing, managemental procedures and disease control measures, poultry farming can be utilized as a profitable source of income not only by the marginal farmers of Northeast Region but for the unemployed youths also.

Some Important Points in Poultry Rearing:

Principles of poultry feeding:

Poultry feeding is one of the important branches of poultry Science, since feed cost alone accounts for 70% of the total farm expenses. It is in the interest of every poultryman to be able to get as much chicken meat or as many eggs for every rupee that he/she spends on the feed as possible.

The problems involved in the feeding of fowl in recent times are multifarious because of modern farming conditions such as: (i) increased cost of poultry feed (ii) confinement rearing of poultry (iii) inclusion of agro-industrial by-products in poultry rations and (iv) increase in flock size and high flock density.

While computing ration for poultry birds, the following facts should be considered:

i. Poultry birds have no lips or teeth, hence require more concentrated ration.

ii. Their digestive tract, having a simple stomach is comparatively short, digestion is quite rapid. It takes about 2½ hours for feed to go from mouth to cloaca in the laying hen, and 10 hours in a non-laying hen. Therefore, the nutritive requirements of poultry are more precise.

iii. Unlike ruminants, where microorganisms synthesize a sizeable portion of essential amino acids, vitamin B-complex, vitamin K in the stomach, the poultry completely depend upon the dietary source for all such nutrients (all essential amino acids, vitamin B complex, etc)

iv. Poultry birds are fed collectively rather than individuals.

v. Due to higher rate of metabolism poultry require a more exact ration.

Raising of Broilers:

A broiler or fryer is a young chicken of either sex below 6 weeks of age weighing 1.5 to 2.0 kgs body weight, with a tender meat, soft, pliable smooth textured and flexible breast. Roaster on the other hand is also a young chicken but is much older (12 to 16) and heavier than broilers. It may be of either sex but tender meat with soft, pliable, smooth textured skin and breastbone cartilage, which are somewhat less flexible than of a broiler or fryer.

To get quick returns, broiler raising is now playing an important role in rural economy as one flock is ready for marketing within 1½ month.

Broiler Feeds:

Since feed constitutes about 70% of the cost of producing broilers it is important to give special attention on it. Broilers are usually fed with two types of rations, broiler starter and broiler finisher. Later trend is to feed the former ration up to 4weeks of age and the later is continued till the age of marketing (6 or 7 weeks whatever it may be). It may be noted that starter ration is having more protein (21-22%) and less metabolisable energy, 2,900 Kcal than the finisher diet, containing increased levels of fat and xanthophyl pigments, aids in the development of the uniform yellow skin colour.

Broiler Health Programme:

A health programme is fundamental to successful broiler production. A suggested disease prevention and control programme follows:

i. Start with disease free chick
ii. Vaccine chicks against Rankhet and Marek's disease at the harchery
iii. Use effective drugs in the feed, or a vaccination programme to prevent Coccidiosis.
iv. Keep feed from aflatoxin.

v. Do not allow visitors or attendants inside the broiler house unless they wear disinfected boosts and clean clothing.
vi. When there are several age groups on the farm, always care for the youngest birds first while performing daily routine works.
vii. Rework built-up litter. When built up liter is used, all caked and wet litter should be remove and replaced with fresh, clean litter before arrival of new chicks.
viii. Cover floor with clean litter at 3 inch deep after each clean out, wood shavings, rice huils, straws cut into small pieces are suitable litter materials.

Common Deficiencies in Broiler Farm Management:

Here is a list of the most common management faults found in broiler houses:

i. Dirt water troughs
ii. Insufficient water space
iii. Insufficient feeder space
iv. Insufficient litter in nests-dirty eggs.
v. Improper ventilation-houses too tight, usually. This is a very common deficiency in developing industries.
vi. Feeders too full or improperly adjusted so that feed is wasted
vii. Tendency to give too much medication. Constant medication is an indication that the manager is using drugs as a substitute for basic management.
viii. Wrong medication due to failure to obtain competent diagnosis.
ix. Allowing weeds or tall crops to block air flow from the poultry buildings.
x. Allowing lice and mites to build up enough to cause production drops due to lack of parasite inspections.
xi. Failure to follow a vaccination programme, many flock owners have problems with flocks that have not been vaccinated properly with even the most common vaccines.
xii. Caked or damp litter in the poultry house

xiii. Cannibalism is cause by crowding, lack of feed or water space or improper ventilation.

xiv. Insufficient grit and shell hoppers when self-feeding of calcium is used.

xv. Failure to keep feed consumption, mortality and production records.

xvi. Failure to follow advice of better service person.

If the above items are taken into consideration and eliminated as deficiencies, the production of farm will increase and the farmers will be benefited.

Reference:

World's Poultry Science Association 2009, World's Poultry Science Journal, Vol. 65, March 2009

GUÈYE, E.F. and VAN'T HOOFT, K. (2002) Networking for family poultry development. LEISA Magazine18 (1): 36.

HUQUE, Q.M.E. (1996) Improving skills of the small farmers in poultry management, Proceedings 20thWorld's Poultry Congress, New Delhi, India, Vol. 1, pp. 47-60.

Information dissemination for family poultry: E.F. GuèyeWorld's Poultry Science Journal, Vol. 65, March 2009 123

Banerjee G.C,Poultry,Oxford & IBH Publishing Company, New Delhi.

* * *

15

LIVESTOCK PRODUCTION AND WOMEN-A COMPATIBLE PAIR

Smt. Ajanta Deb (Kar)
(HOD) Dept. Of Economics, Don Bosco College, Tura
E-mail: ajantakardeb@yahoo.com

ABSTRACT

Man and animal were together since time immemorial. In India, livestock is an important subsector of agriculture and it plays an important role in the national economy. In the success story of livestock sector in India, major contribution comes from the rural women, who are mostly illiterate. Livestock sector is an important source of employment in rural India, particularly for women. Women work in livestock in a better way than men. Many of the important tasks in animal husbandry activities are performed by women besides fulfilling their responsibilities as home makers. Role and involvement of women in livestock activities vary with different livestock production system all over the world. Women play an important role in activities dealing with livestock management, transformation and marketing. In view of raising productivity of livestock sector, women need to be empowered through training, greater access to and control over resources and equal participation with men in decision making.

Keywords: Livestock, Women, Empowerment, Joint responsibility

Introduction

A bird cannot fly with one of its wings only. Similarly, a nation is able to march forward when it is well equipped with quality manpower—both men and women. It is in this context that we intend to study the position held by women in Livestock sector, who otherwise, usually, take a backstage when it comes to the socio economic status of poor, illiterate women, particularly in the developing Third world countries.

Since dawn of civilization, men and animal were together. Livestock is considered to be the oldest wealth resource for mankind and was a status symbol for man in the society. The term 'Livestock' in its broader perspective, refers to any breed or population of animal kept by humans for a useful, commercial purpose. Livestock is also defined as a variety of capital, explained from various angles:

- Livestock per se is a natural capital that has shaped the lifestyle of myriads of human communities with its products and services through time all over the world.
- Livestock is also a financial capital as it can be a primary source of saving, income, credit insurance, loans, gifts and investments.
- Livestock is also used as a productive physical capital. For pastoral societies it is the 'vital' technology that allows producing valuable products from limited resources. Livestock can store and transfer food from one season to another, thus buffering critical seasonal food gaps, and its transport capabilities allow for pastoral mobility and trade patters.
- Livestock helps generate social capital by traditionally providing employment, wealth, prestige, identity, respect and connectedness within and outside the community.
- Livestock also plays an important role as human capital by enabling more healthy lifestyles and empowering people, particularly women.

Livestock system is a land-use unit comprising of animals and auxiliary feed sources transforming plant biomass into animal products. (Panda, 2006)

Livestock Production System Based on Gender:

India is rich in agro- ecological diversity. It has helped in developing a range of unique Livestock production systems that have evolved in each region in tune with the naturally available resources and needs of the people. In India, the two most common production systems found across rain fed agricultural zones, are i) mixed crop-livestock farming and ii) pastoralism. In mixed crop-farming, farmers derive their livelihood somewhat equally from agriculture and livestock; in pastoralism, people's livelihood depends primarily upon their livestock, which are exclusively maintained on grazing.

The livestock production systems can be broadly divided into four categories of increasing responsibility based on gender and age:

i) No involvement by women
ii) Women responsible only for processing products
iii) Women responsible for managing and processing small and other animals kept at the homestead
iv) Women responsible for managing and herding large stock and other animals and for the processing of livestock produce.

Role of Women in Different Livestock Production System:

Role and involvement of women in livestock activities vary with different livestock production system all over the world. In general, livestock production system can be divided into four categories (Niamir, 1994).

i) Transhumant
ii) Agro pastoral
iii) Intensive crop and livestock
iv) Peri-urban intensive systems

Almost all women in transhumant societies own poultry, guinea fowl and other small stock. The major characteristic feature of this system is where women and girls are responsible only for procuring, processing and marketing livestock products, particularly milk and hides (Hudson, 1966 and Lamphear, 1976), but are rarely assigned major herding and management responsibilities for large stock.

The most common agro pastoral pattern is where women own cattle and small stock, but manage and control only the animals that stay around the homestead. Feeding, herding, vetting, collecting and processing secondary products of livestock are clearly women's work (Deere and Leon de Laal, 1982, Hecht 1983; Finney 1988). Like transhumant women, agro pastoral women almost everywhere are involved in dairy production, and in raising poultry and other small stock.

In Intensive crop and livestock systems, East Asian women generally have more control over livestock ownership than elsewhere and have total responsibility for the care and use of small animals, poultry and donkeys. Here, women are mostly engaged in milking, feeding and cleaning of animals. Caring for sick animals, calving, fodder and hay collection are usually done jointly, by both men and women. Fodder production, the purchase of feed and medicines etc. attending to difficult births, taking the animals for breeding and the care and handling of male animals etc, are usually done by men. In terms of decision making, the handling and marketing of milk is mostly done by women. Decisions on the sale of female animals are taken by both men and women (Rangnekar, 1991).

In Peri- urban systems, urban people take to livestock activity to supplement their incomes (Baker and Pederson, 1993). This system is growing in importance, at least in Africa and Middle East. According to one evaluation, women's Peri-urban farms are generally much bigger than those of men (FAO 1983). In Iran, the Peri-urban livestock sector is controlled by commercial elite rather than by the middle class or urban poor; who are basically livestock traders, butchers or large breeding herds(Sanford 1977a).

The roles of women in livestock production in the developing world are as diverse as the women's ethnicity. Evidence from around the world shows that pastoral women's rights to property and assets are being increasingly eroded, undermining their status in relation to both the household and the wider society. The reasons behind this may be cited as

i) Commercialization
ii) Sedenterisation
iii) Labour migration and
iv) The internal breakdown of the socio-economic system.

Women play an important role in activities dealing with livestock management, transformation and marketing. A vast majority of livestock producers comes from under privileged section of rural community and needs a livestock development and research paradigm to achieve sustainable livestock development. Studies have shown that development of small holders' mixed-crop livestock production is one of the most effective methods of poverty alleviation. In rural sector, poverty alleviation programme should include dairy sector as its base as it has the potential to bring about maximum growth with minimum capital outlay. (Rupinder Kaur 2008). The output in dairy sector increased by over four times and its share in total value of output from livestock sector increased from about 55% in 1951-52 to about 67% in 2003-04. The dairy sector contributes the largest share in Agricultural GDP of our country. The value of milk group was higher than the value of output from rice, the largest contributor to agriculture amongst food grains. In general, women play a predominant role in dairy farming; compared to men, more women participate in dairy work. Women account for 93% of total employment in dairy production in India. But women are under recognized when it comes to formulating policies w. r. t provision of inputs including credit, dissemination of information related to modern dairy practices and inclusion in the training and extension programmes. To raise women's active participation in livestock activity, the following points have to be taken note of:

- Credit facility should be extended to the doorstep of women; micro finance module may be more practicable in this respect.

- Strengthening of co-operatives with more women representation, having more access and control over their earnings; more women societies should be encouraged.
- Availability of right information and training is a pre-requisite in this regard.

Benefits of Marginalized Women from Better Access to Land and Livestock:

Access, control, and management of resources such as grazing areas and feed resource, provide assets that improve women's equality and empowerment with an overall positive impact on the welfare of the household.

- Land and livestock ownership is considered to increase a woman's decision making power inside the household, and a better control over the income from the sale of livestock produce inside the village.
- Both land and livestock can provide better security for elderly women who are often solely dependant on their families. These assets become her bargaining tools for better care from relatives in old age.
- Ownership and access to land and livestock can generate more income than most of the other economic activities; marginalized women are currently involved in.
- This can help some widows unwilling to remarry and support her children, without being burden to others.
- Ownership of land and livestock can open up possibilities of accessing credit.

  Some of the common constraints that the marginalized women face all over the developing world are:
- Culture and tradition
- Lack of credit
- Lack of land
- Lack of shelter for livestock and
- Poverty

Empowering Women through Training and Extension Programme:

Various surveys show that women in all countries are interested in learning new techniques and ideas. The major change that has occurred in the livestock sector is the participation of women in training programmes. Micro level Studies carried out with women to understand their views and suggestions about training indicate that the conventional approach to training has to be radically changed. The following major aspects regarding women's perceptions and suggestions on training need to be kept in mind while planning training programmes for women:

- Women basically dislike long lectures and can effectively learn while doing; hence the programme should be practical.
- Women prefer discussing problems that they currently face.
- Illiteracy is very high amongst rural women and long notes are of limited use. Audio-visual material should be used to the maximum extent.
- The women suggested short-term training covering few subjects.
- They suggested training or meetings to take place in the afternoon, when they are relatively free.
- Factors like socio-economic status, age etc. have a bearing on interaction and active participation of the women.
- Use of local dialects is strongly recommended.
- Involvement of women training and extension officers would increase effectiveness of the programme.

Though rural women in our country are engaged in a Herculean task of managing the household, attending farm labour, looking after domestic animals, etc. still they do not enjoy enough decision making power, particularly in financial money matters. Empowerment of rural women is a necessary component in the advancement of livestock sector in India.

Empowerment as a Goal:

Usually empowerment for women involves opening up of more opportunities, greater access to and control over resources and equal participation with men in decision making, which are often cited as 'strategic needs' as opposed to 'basic needs' for survival. Tactics for empowerment or addressing of such strategic needs can involve increasing:

i) Awareness
ii) Self-confidence
iii) Knowledge and collective activities
iv) Reduced vulnerability to violence and exploitation
v) Economic security, independence, options and opportunities
vi) Shared responsibility for productive work with men and the state
vii) Organizing with other women for strength, solidarity and action
viii) Political power
ix) Ability to improve the lives and futures of their children and
x) Humanistic and just development processes.

Various dimensions of Empowerment of Women can be put under the following heads:

- Psychological: these are basically self confidence, creating space, acquiring knowledge;
- Social: these are community action, action for rights, social inclusion, and literacy;
- Economic: these are income security, entrepreneurial skill, ownership of productive assts;
- Cultural: these are redefining cultural rules and norms (ethnic, gender), recreating cultural practices.

In many developing countries, women provide much of the labour for livestock tasks. Yet their role in livestock production has been undervalued by policy makers and research on this issue is widely ignored. In many countries, women are often denied ownership rights for large stock (cattle, camels, horses), but 'allowed' to keep small stock (sheep, goats, rabbits, poultry). Reasons are varied:

i) Livestock patterns are linked to social class, religious systems and paternalistic cultures, which mean women, have weaker ownership rights compared to men, especially in times of stress.
ii) The migration of men to find seasonal work makes it harder for women in their households to use land, or access credit and technical inputs for livestock production.

Livestock sector plays an important role in Indian economy and is an important sub sector of Indian agriculture. India has the largest livestock population in the world. Distributed over 100 million households in approximately 600,000 villages, Indian farmers have a variety of livestock -ranging from little known Yak and Mithun to the apparently insignificant backward poultry. India has a livestock population of 485 million, a poultry population of 489 million, (Livestock census - 2003) – being the first in cattle and buffalo population, second with respect to goat and third in sheep population in the world. India has 57% of the world's buffalo and 16% of the world's cattle population.

The contribution of Livestock sector in India's Gross Domestic Product (GDP) was 4.7% in 2004-05 at 1999-2000 prices. The overall growth rate in the Livestock sector is more or less steady, hovering around 4-5%, though investment in this sector is never substantial. According to CSO estimates, the share of Livestock sector in the GDP of Agriculture and allied activities was 24% (1993-94 prices) during 1999-2000, which rose to 27% in 2003-04.Poultry, which was considered as a backyard venture in the early 60s, has now been transformed into a major farming activity. India occupies 4th position in the world with respect to egg production, which has reached to 45.2 billion (2004-05). In milk production India is leading the world and has reached a level of 91 million tones in 2004-05.

In this success story, major contribution comes from the rural women, who are mostly illiterate, consisting of 71% of labour force in livestock farming; in the dairy sub sector of Livestock, about 75 million women are employed as against 15 million men. (Govt. of India, 2001).The rural women play a significant role in Animal husbandry and are directly involved in most of the operations relating to feeding, breeding, management and health care of the Livestock. As per 1999- 2000 NSS data, percentage distribution of male and female workers in rural areas in Animal husbandry is 49.6% and 50.4% respectively (Srivastava and Srivastava, 2010).

In rural areas, about 83% women workers were engaged in agriculture in 2004-05 either as cultivators or laborers as compared to 67% male workers. (Srivastava and Srivastava, 2010). In India, the Livestock continues to be raised on crop residues and agricultural bi-products. The area under cultivated fodder production is limited only to 4.6% of the total cultivable land. The ownership of the Livestock is more evenly distributed with landless labourers and marginal farmers, majority of who are women, owning bulk of Livestock.

In India, the Livestock production and agriculture are intrinsically linked, each one being dependent on the other and both crucial for the overall food security of the people. Livestock sector is an important sub sector of Indian agriculture and it forms an important livelihood activity for most of the farmers, supporting agriculture in the form of critical inputs, contributing to the health and nutrition of the household, supplementing incomes, offering employment opportunities, particularly to the landless, small, marginal farmers and the women. Livestock provides stability to family income especially in the arid and semi- arid regions of the country. In the arid areas, the contribution of Livestock to Agricultural GDP is as high as 70% while in semi- arid areas the contribution is over 40%. Besides contributing food and being important inputs for crop production, Livestock are important as savings or investment for the poor households and provide security or insurance through various ways in different production systems. (Kitalyi *et al,* 2005). Livestock manure is the major source of nutrients for organic crop production and for sustaining soil fertility. Livestock wealth is more equitably distributed than that of land and hence become more important for the poorer household, in particular. It is unfortunate that despite legislative changes, few women have control over land (Srivastava *et al*, 2008). Moreover, Livestock rearing contributes to on-farm diversification and intensification, which could be one of the strategies for poor household to escape poverty and to maintain some stability in their earnings.

Livestock sector is an important source of employment in rural India, particularly for women, with a participation rate of as high as 71% as compared to 33% in crop farming. Women play a major role in Livestock farming, especially small ruminants and poultry, with a good knowledge of Livestock behaviour and local feeds. Female headed households are more likely to rear small ruminants and poultry as they depend more on common property resources for meeting their fodder and feed requirement. (Kumar *et al*, 2008).Animal husbandry increases the earning capacity of women and ultimately leads to their economic empowerment. Of late, an increasing tendency is being noticed regarding women's participation in Livestock rearing activities. One main reason behind this is the continuous migration of men from rural areas to towns and cities looking for greener pastures. This has created more households with female heads who puts a lot of inputs in the form of their day to day activities for economic returns like sale of Livestock and Livestock products or using of Livestock in various livelihood activities.

Traditionally, women are involved in agriculture and Livestock production activities in rural areas (Waters-Bayers, 1985). Reports are there from many parts of the world that women work in Livestock in a better way than men (Ishani, 2004). Animal husbandry is predominantly a male affair in case of farmers with high economic status as majority of them employ permanent male labour to look after the animals, whereas it is predominantly a female affair in case of farmers of medium and low socio-economic status. (Verma,1992). Many of the important tasks in animal husbandry activities are performed by women besides fulfilling their responsibilities as home makers (Randhawa and Chandra, 1993). On an average, a woman devotes 3.5 hours per day for animal husbandry activities as against only 1.6 hours per day devoted by men in this category. There is a distinct sphere of participation between males and females in farm and dairy sector, the men were the planners and women implemented the activity (Sangwan et.al, 1990).

Livestock sector is unique in terms of employment opportunity as two-third of female workforce in rural India is engaged in livestock rearing. (Kumar *et al*, 2008). Though the average holding of livestock is small, the livestock sector has considerable potential for generating additional employment through milk, meat, and wool and eggs production. Milk production alone involves more than 30 million small producers. Women play a major role in livestock production and most of them have good knowledge about livestock behaviour and local feeds. In order to raise their efficiency and also productivity of Livestock sector, women should be given adequate training in marketing, particularly cooperative marketing. They need to have better accessibility to market information, controlling the means of production and in gaining access to transport and storage facilities for their produce.

Despite their considerable involvement and contribution, women's role in livestock production has been underestimated, undervalued and widely ignored. This is due to a paternalistic bias from men but also from women, who are often conditioned by their culture and society to undervalue their own worth. In addition, information on the role of women in livestock production often obscures the actual contribution of women, because of the biases of the respondents, the biases of surveyors and biases inherent in the definitions of employment, e.g., informal versus formal, or establishing a minimum number of hours of work to qualify as economically active (Morton 1990). Women's role in agriculture and livestock production is better documented in sub-Saharan Africa than elsewhere (Safilios-Rothschild 1983).

Rural development becomes complete only when along with Agricultural growth policy, other aspects like animal husbandry are focussed for its due share. Operation Flood is an example of higher women participation which reminds us of our joint responsibility and strengthening our rural women folk for our own benefit.

References:

Baker, J. & P.O. Pedersen, eds. 1993. *The rural-urban interface in africa: expansion and adaptation.* Seminar Proceedings No.27. Scandinavian Institute of African Studies, Uppsala

Deere, C.D. & M. Leon de Laal. 1982. *Women in Andean agriculture.*] ILO, Geneva

Dyson-Hudson, N. 1966. *Karimojong politics.* Oxford: Clarendon Press

FAO. 1983. *Expert consultation on women in food production.* ESH Division, FAO, Rome

Finney, R. 1988. The role of women in the adoption of new practices. In *Proceedings of the sixth world conference on animal production,* WAAP Helsinki, 213-31

Hecht, S.B. 1983. Environmental effect of conversion of tropical forest to pasture in the Eastern Amazon. In *The dilemma of Amazon development,* ed. E. Moran. Boulder, Colo.: Westview Press

Kaur R (2008) "Gender and Social analysis of Dairy farming: A case of Punjab" Journal of Rural Development 27 (1): 129-147

Kitalyi A, Menga L, Morton J, Mcleod A, Thornton P, Dorward A, and M Saadullah (2005), " Why Keep Livestock if You are Poor? Livestock and Wealth Creation"Nottingham University Press,Nottingham

Kumar A and D. K Singh (2008) "Livestock production systems in India: An Appraisal across Agro-Ecological Regions". Indian Journal of Agricultural Economics 63(4)

Lamphear, J. 1976. *The traditional history of the Jie of Uganda.* Oxford: Clarendon Press

Morton, J. 1990. Aspects of labour in an agro-pastoral economy. The northern Beja of Sudan. Pastoral development Network, Paper 30b, ODI, London

Niamir-Fuller,M.1994. Women Livestock Managers in the Third World: Focus on Technical Issues Related to Gender Roles in Livestock Production.IFAD Staff Working Paper No 18

Panda,S.C 2006 Cropping and Farming Systems.Agrobios (India)

Planning Commission (2007) " Report of the Working Group on Livestock and Animal Husbandry" Government of India, New Delhi- (2008) Eleventh Five Year Plan (2007- 2012) Government of India, New Delhi

Randhawa A and S Chandra (1993) "Changing role of home Science Scientists in transferring farm technologies to farm women in Agriculture" Paper presented at national seminar on " Women in Agriculture- development issues at NAARM" Hyderabad Dec 28-30

Rangnekar, S.D. 1991. *Women in dairy production – observations from India. WilD* (May/June). Heifer Project International, Little Rock, Ark

Safilios-Rothschild, C. 1983. Women in Sheep and goat production and marketing. Expert consultation on women in food production, FAO, Rome

Sanford, S. 1977a. Pastoralism and development in Iran. Pastoral Development Network, Paper 3c, ODI London

Sangwan, V., Munjal, S. And Punia,P.K.1990. Participation of farm women in farm activities. Indian J. Extn.Edu, 26(1and 2):112

Srivastava N and R Srivastava (2010) "Women, Work and outcomes in Rural India" Economic and Political Weekly XIV (28): 49-63

Srivastava R, Saxena N. C and S. K Thorat (2008) "Land Institutions, Policy and Reforms in India" in Ashoke Gulati and Shenggen Fan (ed) The Dragon and the Elephant: Agricultural and Rural Reforms in China and India. New York, John Hopkins Press pp 71-96

Verma S. K (1992) "Women in Agriculture: A Socio Economic Analysis" Concept Publishing Company, Commercial Building, Mohan Garden, New Delhi

Waters-Bayer, A. 1985. Dairying by settled Fulani women in Central Nigeria and some implications for dairy development. Pastoral Development Network, Paper 20c, ODI, London.

* * *

16

LIVESTOCK AND ENVIRONMENTAL POLLUTION

BISWAJIT PAUL* AND B.K.MISHRA**

*Lecturer (Selection Grade), Department of Zoology, Don Bosco College, Tura, E-mail: bptura@yahoo.co.in

**Assistant Professor, Department of RDAP, NEHU,Tura Campus,Meghalaya

ABSTRACT

With the demand for food rising every moment there is an emergent need to compensate the demand. One of the ways to neutralize the ever expanding demand for food is through the rearing of livestock as it not only provides food to the masses but also brings about an economic stability to the poorer section of the society as it is observed that livestock rearing is within the reach of a lower income group household. But this aspect also makes it true that the enlarging global increase in livestock rearing is having an adverse effect on the environment both in production level and also in consumption level. It is with this view this paper is brought out.

Keywords: livestock, population, demand, environment, biodiversity.

Introduction:

During the transition from hunter gatherer to settled farming, human have envisaged the importance of animal rearing. For thousands of years man has been rearing and domesticating the animals which were once in the wild and today the livestock rearing has taken the shape of an industry with a great potentiality to reckon with. Livestock generally means rearing of domestic, semi-domestic or captive wild animals for useful, commercial purposes. Some refers livestock to only domestic animals or even just to red meat animals. Importance of livestock rearing is multifaceted as it provides nutritional components in the form of protein, carbohydrate, fat, vitamin, minerals etc. It also contributes to the products of dairy, primarily milk – a complete natural food and secondarily yoghurt, cheese, butter, ice cream etc. The contribution of livestock to fiber production is well established as it produces wool and mohair. Skins are utilized from cows; sheep for leather and bones, hooves and horns for medicinal and recreational purposes. The horses, yak, donkeys etc. has been utilized as potential labor force from time immemorial. As they form a mechanism of transport, especially in the rural areas, their use in agriculture for ploughing fields are still prevalent in many areas even today as they cheaper labor force compared to the mechanized versions. The Military has also its share of utilizing the livestock in times of need. Livestock manure serves as a good fertilizer. Grazing does help sustain vegetation and promote biodiversity by dispersing seeds, controls shrub growth – especially the dry ones which may lead to wild fires and destroy biodiversity, breaking soil crusts, stimulating grass growth and improving seed germination. Livestock also has its role in savings and currency. Cultural value of livestock in rural India is well known as it plays a role as gifts in marriages and resolve conflicts between families.

So keeping all the above views in mind the production and utilization has become one of the priorities of nations and has now taken the shape of an industry. The demand for food and other agricultural products have risen with the rise in population. Today the world population is at a staggering 6.5 billion and growing at the rate of 76 million annually (UN, 2005). The projected population till 2050 is 9.1 billion and would peak up to 9.5 biliion by 2070 (UN, 2005). The rise in population would demand for an adequate supply of food. This supply would be met by development is agricultural sector, which includes –crop production and livestock production on the other. But it is interesting to note that though the demand have been steep to satisfy the hunger, agricultural production as a whole has remained at a standard rate of 0.3% GDP through out the world. But if livestock production is seen, then the scenario is different where the production has increased manifold. Consumption has a trend to follow in relation to urbanization and economic development. In 2005, 49% of world populations were living in cities (FAO, 2006b). Urbanization is a global phenomenon and all population growth between 2000 and 2050 will be urban (FAO, 2003a). The pattern of food consumption would change along with urbanization as people would prefer more of pre cooked, fast and convenience food, snacks etc. when away from home (Schmidhuber and Shetly, 2005; Rae, 1998; King, Tietyen and Vickner, 2000).

Figure: Past and projected global rural and urban populations from 1950 to 2030.

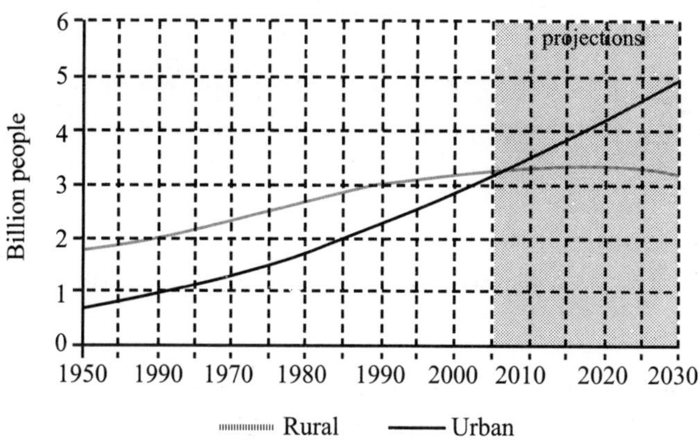

Source: FAO (2006a) and FAO (2006b)

Consumption kg/person/yr						
	1992	1970	1980	1990	2000	2003
Cereals	132	145	159	170	161	156
Roots, tubers	18	19	17	14	15	15
Starchy roots	70	73	63	53	61	61
Meat	10	11	14	19	27	29
Milk	28	29	34	34	45	48

Figure: Past and projected Food consmptn. of LS products

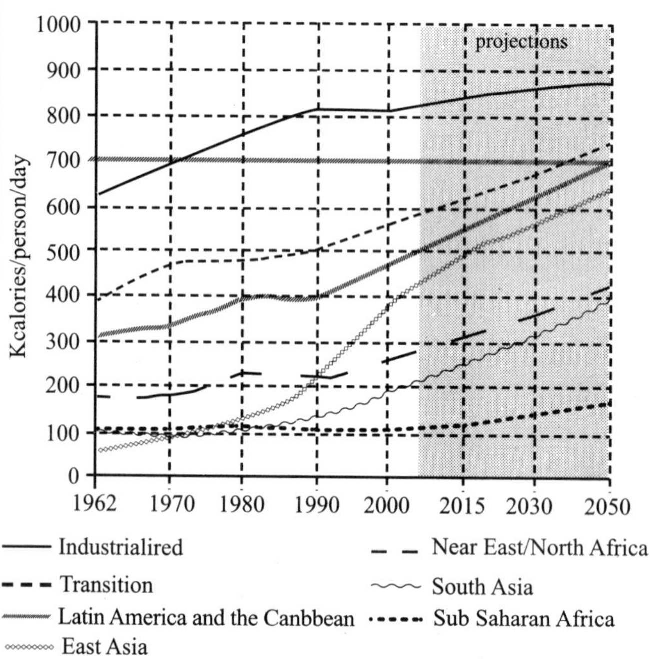

Note: for past, three-year average centered on the indicated year. Livestock projects include meats, eggs, milk and dairy products (excluding butter).

Source: FAO (2006a) and FAO (2006b)

Source: FAO (2006b)

India presides over the worlds' largest democracy, embracing countless cultures, languages and religions, but also a population exceeding one billion- a number to reckon with. India is expected to become the worlds' third largest economy after US and China by 2010 (Jones Lang La Salle, 2005). The UN's "World population Prospects", released in 2005, estimates that there will be 1,393 million people in India by 2025. The per capita income is showing a fast growing trend. Personal disposable income annually is set to increase 8-10% per year over the period 2006-2010. Indian National Council of Applied Economics research expects a further 180 million to join the middle class category by 2010.

So the above parameters i.e., increase in population, urbanization and per capita income has a correlation with the food intake of the population. The pattern of food intake in the recent past has shown a significant change (Kumar and Birthal, 2004). There is a positive decline of Cereals and significant increase of consumption of fruit, vegetables, milk products, meat, fish and eggs. The animal based food is more preferable across the board as it provides high value foods with lesser expenditure and so pro poor.

The table below shows the consumption of milk and meat during the last few years.

Product per year	1996	'97	'98	'99	00	01	02	03	04	AG
Milk	193.04	200.38	206.38	213.95	217.75	224.82	227.87	234.25	233.12	2.301
Meat	13.81	13.23	13.48	13.70.	14.30	14.63	14.88	15.29	15.29	1.293

AG – Average Growth.
Source: FAO. Stats 2006.
Table: Per capita consumption of Livestock Products (gm/day/person)

Along with the above factors the trade liberalizing policy of Indian Government has given a fillip to the livestock industry in India. A number of policy initiatives were taken by Govt. of India to boost export of livestock products, especially buffalo meat (Birthal and Taneja., 2005).

The Livestock production from 1985 to 2005showing a notable growth in the table below:

Year	1985	1990	1995	2000	2005
Milk (million tons)	44.02	53.68	65.25	80.83	91.94
Beef and Buffalo meat (million tons)	1.95	2.4	2.72	2.86	2.98
Sheep and goat meat (million tons)	0.53	0.61	0.66	0.70	0.71
Poultry meat (million tons)	0.19	0.37	0.62	1.14	1.97

Source: FAO STAT, 2006

Livestock population in North East has recorded moderate high growth Bhowmick and Kalita, 2005. Assam has the largest stock of all the livestock animals. The growth per annum of cattle population is about 3.87%. Meat, milk, eggs are the common livestock product is used. National Agriculture Policy (2000) target 4% annual growth rate by 2020. But at present it is much below that target and it is only 2%. Though agricultural sector is not keeping up with the target, the livestock sector is contributing handsomely to the GDP. Sustained Economic growth, rising income and urbanization are driving the livestock sector forward keeping parity with the rest of the world.

But though large scale livestock production units are in a position to cope up with the rising demand in the agricultural sector its contribution to the environment is not sound and rather posing a threat to the development. Livestock is affecting all the avenues of environment and causing health hazards to animals and human alike. Their pressures on Biodiversity are a potential threat.

Effect of Livestock production on air:

Piling up of too much stuff in one place, causes problems. If you spread out the animals and let manure lay where it falls in a pasture, it doesn't bother any one very much. But if you start collecting it, flushing it, spreading and spraying around – all normal practices in confinement operations – it becomes air pollution."

Spray applications of liquefied manure to fields is a method to dispose wastes of livestock. In this process, large amounts of manure sends dust particles into the air that penetrate into the lungs of human and other animals. The particles carry toxic gases such as ammonia (NH_3) and Hydrogen sulfide (H_2S). NH_3 can impede the lungs from clearing dust particles and H_2S can prevent cells from using Oxygen (O_2) and cause loss of consciousness, coma or even death. Most of the effect through air in regards to livestock production occurs subtly but one aspect of Livestock production and manure utilization in regards to air pollution is not subtle and that is 'odour'. Chronic bad odour has significant bad effects. Global anthropogenic emission of NH_3 has recently been estimated at some 47 million tons N (Galloway et al., 2004). And livestock production typically contributes 70 to more than total NH_3 emissions by anthropogenic means (Bouwman et. al., 1997). Green house gases (GHG) like Methane ($CH\$_4$) and Nitrous oxide ($N_2O$) are emitted during livestock manure management. Livestock accounts for about 35 – 40 % of the total anthropogenic methane emissions. The most important source of methane emission in India is enteric fermentation from domestic livestock, which is showing a steady increase over a period of time. During 1980-81 it was 9820 Gg and increased to 11,790 Gg during 1997-1998. The share of enteric fermentation from Livestock was more than 68% of the total methane emissions from the agriculture sector during 1996-1997.

Past and current concentration of important greenhouse gases

Gas	Pre-industrial concentration (1750)	Current tropospheric concentration	Global warming potential*
Carbon dioxide (CO_2)	277 ppm	382 ppm	1
Methane (CH_4)	600 ppb	1728 ppb	23
Nitrous oxide (N_2O)	270-290 ppb	318 ppb	296

Note: ppm = parts per million; ppb = parts per billion; ppt = parts per trillion; *Direct global warming potential (GWP) relative to CO_2 for a 100 year time horizon. GWPs are a simple way to compare the potency of various greenhouse gases. The GWP of a gas depends not only on the capacity to absorb and reemit radiation but also on how long the effect lasts. Gas molecules gradually dissociate or react with other atmospheric compounds to form new molecules with different radiative properties.

Source: WRI (2003); 2005 CO_2; NOAA (2006); GWPs: IPCC (2001b)

Livestock activities contribute substantially to the emissions of N_2O – the most potent out of the three major green house gases. Livestock produce about two third of all anthropogenic N_2O and 75 – 80 % of agricultural emission.

Table: Role of livestock in carbon dioxide, methane and nitrous oxide emissions.

Gas	Source	Mainly related to extensive systems (10' tonnes CO_2 eq.)	Mainly related to intensive systems (10' tonnes CO_2 eq.)	Percentage contribution to total animal food GHG emissions
CO_2	**Total anthropogenic CO_2 emissions**		**24 (-31)**	
	Total from livestock activities	-0161-2.71		
	N fertilizer production		0.04	0.6
	on farm fossil fuel, feed		-0.06	0.8
	on farm fossil fuel, livestock-related		-0.03	0.4
	deforestation	(-1.7)	(-0.7)	34
	cultivated soils, tillage		(-0.02	0.3
	cultivated soils, liming		(0.01)	0.1
	desertification of pasture		(-0.1)	1.4

	processing		0.01-0.05	0.4
	transport	-0.001		
CH_4	**Total anthropogenic CH_4 emissions**	**5.9**		
	Total from livestock activities	2.2		
	enteric fermentation	1.6	0.20	25
	manure management	0.17	0.20	5.2
H_2O	**Total anthropogenic N_2O emissions**	**3.4**		
	Total from livestock activities	2.2		
	N fertilizer application		-0.1	14
	indirect fertilizer emission		-0.1	1.4
	leguminous feed cropping		-0.2	2.8
	manure management	0.24	0.09	4.6
	manure application/deposition	0.67	0.17	12
	indirect manure emission	-0.48	-0.14	8.7
Grand total of anthropogenic emissions		**33 (-40)**		
Total emissions from livestock activities		**-4.6 (-7.1)**		
Total extensive vs. intensive livestock system emissions		**3.2 (-5.0)**	**1.4 (-2.1)**	
Percentage of total anthropogenic emissions		10 (-13%)	4(-5%)	

Effect of Livestock Production on Soil:

Land degradation is widely recognized as a global problem having implications for agronomic productivity and the environment as well as effects on food security and quality of life (Eswaran, Lal and Reich, 2001). The magnitude of livestock associated land degradation is not yet established in detail because it is difficult to quantify the degradation as because of the complexity and lack of easily measurable land quality parameters. But it is certain that if grazing areas are overused it would modify the physical and hydraulic properties of soils and ecosystems. The soils infiltration capacity and organic matter is reduced when vegetation cover declines. Porosity and level of infiltration is reduced because of trampling which compacts the soil. In water joints the density of animals are more and in these areas the compaction is more. The grazing animals remove the small plants and thereby act as a cog in the succession process, as the phytomass is reduced by overgrazing. Carbon in the soil, escape as Carbon dioxide. The soil which gets polluted may than lead to water pollution. Overgrazing of land reduces the density of vegetation and the amount of organic matter generated. This, in turn, increases soil erosion from wind and water and decreases soil fertility through loss of nutrients. In arid and semi arid areas these impacts may result to desertification. Riparian areas along streams and lakes are common place of grazing and these areas - hot spots of biodiversity – are degraded by overgrazing. Vegetation is altered or destroyed to accommodate livestock for grazing and this process leads to the degradation of the soil quality. Erosion occurs and the soil is rendered useless for crop production. Livestock sector is a major land user, spanning more than 3.9 billion hectares, representing more than 30% of the world's surface area.

A chart is shown below to depict the global trends in land use area for livestock production and total production of meat and milk.

Figure: Global trends in land-use area for livestock production and total production of meat and milk.

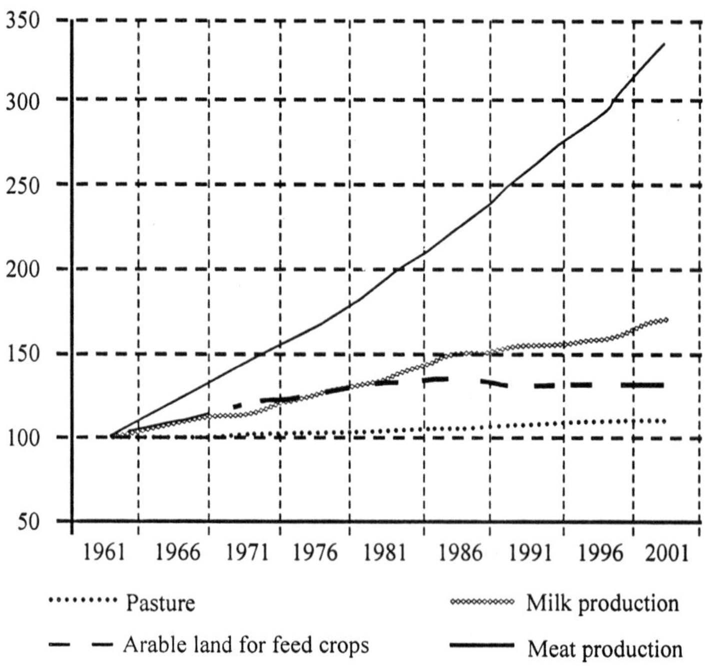

Source: FAO (2006b)

It is estimated that about 100 million cow units graze in forests against a capacity fro 31 million. There is a decline in the area under permanent pastures and grazing land from 1950 onwards because of privatizations, encroachment, distribution of land by the government to the poor, conversion of land to National Parks and Sanctuaries etc (Jodha, 1992). The grazing intensity in India is very high. In rain fed areas, the present stocking rate is 1-5 adult cattle units (ACU) per Ha against the permitted rate of 1 ACU/ha, while in arid zones, the stocking ratio rate are 1-4ACU/ ha as against 0.2-0.4ACU/ha (Shankar and Gupta, 1992)

Effect of Livestock Production on water:

Livestock production sometimes becomes a drain on the natural resources. Animals need more water than grain to produce same amount of food. A sheep fed on grass requires 10 cubic meters of water per kg body weight where as a kg of cereals need only 0.4 to 3.0 cubic meters of water for its growth (Info resources- Focus – 2006). It is also found that 1900 to 4,600 liters of virtual water (total volume of water directly or indirectly required to produce a commodity) was used to produce a liter of milk (Singh et. al., 2004). The pollution of water bodies may be brought by different agencies – like industrial effluents, mining, slurry and outputs from tanneries and other factories, household effluents and also by livestock. The impacts of livestock on water are often under estimated and ignored. Livestock not only brings about a stress on the water by pollution (by waste and Manure) but also by depleting the water quantitatively by consumption. (Roost et al, 2003; Gleick, 2000). But it is understood that degradation is imminent because of the effluents that are discharged from farms either in the form of excreta or in the form of chemicals that are used to keep the farm. The pollution could affect the biota of the water body or even change the structure (physical) of the water body. In both the forms the damage is caused and the effected ones are those who is utilizing that water body and in the downstream. Effect of Livestock on water bodies could be either in the form of Eutrophication, toxification or structural (mechanical) change of the water body. These changes bring about a change in the ecosystem of the water body. Eutrophication is brought about by loads of nutrients in the form of Nitrogen (N) and Phosphorous (P) and organic matter. These nutrients in excessive amounts leads to growth of phytoplankton which leads to the choking of the water body as there is a decrease in water quality and biodiversity, O_2 depletion, fish kills and associated problems along with algal blooms. (Rabalais, 2004). Eutrophication also leads to the blooming of Cyanobacteria – a dangerous toxic bacteria which can take the life of human and other animals when water contaminated with it is taken.

The parameters that are normally taken to ascertain the pollution in water caused by livestock is BOD (Biological oxygen demand) and bacterial contamination. Rise in BOD would lead to the elimination of fish and other aquatic organisms like macro invertebrates, mollusks, amphibians (tadpoles) etc. which are a part of the aquatic system. Three groups of bacteria are indicators of livestock pollution: The Coliform group, fecal coliform bacteria and fecal streptococci (Meehar and Platts, 1978). Toxicological effects are brought about by the Nitrates leaching to ground water aquifers which lead to the birth of blue baby disease (Methemoglobinimia). These nitrates may also penetrate the ground water through leakages of the manure storage systems (Oenema et. al., 2007). Most industrial operations store waste in massive lagoons that can hold millions of gallons of liquid manure. When this manure are dumped in the neighboring areas or when there is a leakage in these lagoons then there is a potential danger of toxicity of the air, water and soil. Antibiotics and Artificial Growth Hormones are commonly used in Industrial farms, either injected directly to the livestock or added to their feed. These substances are excreted and then get in to the water bodies. The hormones are significantly present even for up to 270 days after excretion and are discovered miles down stream from farms (Soto et. al., 2004). These hormones are in minute quantities and may not affect the human but the fishes and the other aquatic organisms would certainly get affected. Antibiotics administered in heavy doses to the animals are also excreted and gets into the water bodies. These antibiotics then produce the dangerous antibiotic resistant bacteria in the water. Humans do get affected by drinking this contaminated water (UCS, 2001).

Effect of Livestock Production on Biodiversity:

Biodiversity refers to the variety of genes, species and ecosystem that can be found in the environment. This diversity is at threat because of various kinds of anthropogenic activities like deforestation, land degradation, mining, jhum cultivation, encroachment of reserves by human et. al. Livestock has also its role in the degradation of biodiversity both directly and indirectly. The role played by livestock to bring about a loss in biodiversity is through loss of habitat, climate change, invasive alien species, overexploitation and competition, pollution.

Livestock are one of the major drivers of habitat change (deforestation, destruction of riparian forests, drainage wetlands) either by itself or by production of feed for it. For convenience of livestock grazing, habitats are degraded and that leads to diversity loss in the form of gene pool, species and ecosystem. In Northern Temperate Regions such as Europe, native vegetation has been largely destroyed or modified by deforestation, agriculture and grazing (Heywood, 1989). Climate change has an effect on biodiversity. And Livestock plays a role to bring about a change in the climate because of emissions of green house gases like carbon dioxide, methane and nitrous oxide. Livestock becomes invasive when it when it encroaches in to a territory where it does not belong. This invasiveness may be brought about by itself or by anthropogenic activities. When it is by itself it interrupts the indigenous species in the wild. But when it is brought by the human then they become a part of the domestic environment and mingle with the local habitat. Overexploitation and competition takes place by three distinct processes: Competition with wildlife leading to reduction of wild life population, Overexploitation of living resources (mainly fish) for use as feed for livestock. And erosion of livestock diversity itself by inbreeding and crossbreeding to maintain good profitable breeds.

Pollution of environment by livestock is diverse as has been discussed previously.

Having said this, we come to a potent question that in regard to livestock production what is happening in the North East India, so far environmental degradation is concerned. It is well documented that all the seven sister states are in some way or the other maintaining a good livestock to meet the need and demands of the market. But what is not recorded is how much damage the livestock production is doing in different states as it has been discussed that the damage gets unnoticed at the initial stages and when noticed then already the damage has been done. So the need of the hour is to check in on the different livestock producing centers if they are maintaining the legal standards and considering to safe guard the environmental pollution.

A case in hand shall substantiate that if necessary steps in due time are taken then disasters can be averted.

Impact of 'ban on grazing' at the Barsey Rhododendron Sanctuary, Sikkim

The Barsey Rhododendron Sanctuary (BRS) lies in the southwest corner of the state of Sikkim spreading over 104km^2 with an altitudinal gradient of 2200m – 4100m. During the year 2000, high impact pastoralism was practiced by 288 households in this sanctuary. They owned 5,370 cows, 370 buffaloes, 506 yaks and 135 sheep which grazed within the sanctuary throughout the year. The livestock density was 61 livestock / km2 and the cattle shed density 4.5 / km 2. Due to trampling by the cattle, 576 ha of land was made barren. Collection of close to 42,000 pole-sized timbers and 22 ton of firewood and 26 ton of fodder was noted from BRS every year. Because of this biomass collection, it leaded to degradation of forest and catchments of the hill streams within that. Not only this was illegal following the Wildlife (Protection) Act of India, 1972, this grazing was also causing a lot of ecological damage in the sanctuary. In order to tackle this, the state forest department along with NGOs like The Mountain Institute and the members of the Eco-Development Committees started phasing out of the cattle and the shepherds from the BRS. By 2005, 276 of these herders (272 cow herders, 1 yak herder and 3 sheep herders) have phased out from this profession. Of the 12 herders still remaining, 10 are yak herders and 2 are cow herders. Of the total 6,324 livestock units grazing within the sanctuary in the year 2000, there has been a 93% reduction and by 2005 only 463 livestock units remain. The livestock density has reduced by 96% from 61 livestock / km 2 in 2000 to 4.45 livestock / km 2 in 2005. Similarly the cattle shed density has also reduced by 96% from 5.54 cattle shed / km 2 in 2000 to 0.23 cattle shed / km 2 in 2005. As the figures indicate a cattle sanctuary has been now slowly converted in to a safe haven for wildlife. There has been 93% reduction in the intensity of livestock grazing within the sanctuary from 2000- 2005. The positive impacts on the wildlife habitats of BRS by the herders is better natural regeneration of the forests, more biomass availability, rejuvenation of the springs and increase in wildlife populations. There has been revival of many mountain springs which currently meets the drinking water needs for many villages adjoining BRS.

References:

Bhowmik, B.C., and Kalita, D.C., 2005. *Department of Agricultural Economics,* Assam Agricultural University, Jorhat, Assam.

Birthal, P.S. and Taneja, V.K., 2006. Livestock sector in India: Opportunities and Challenges for small holders: *Workshop on small holders. Livestock production of Indi.: Opportunities and Challenges.* 31st Jan to 1st Feb 2006. N. Delhi.

Bouwman, A.F., Lee, D.S., Asman, W.A.H., Denteney, F.J., Van der Hock, K.W., and Olivier, J.G.J., 1997. A global high resolution emission inventory of ammonia. *Global Biogeochem cycles,* 11:561-587.

Delgado, C., Rosegrant, M., Steinfield, W., Ehui, S., and Curbois, C., 1999. Livestock to 2020: The next food revolution. *Food, agriculture and the environment. Discussion paper 28.* Washington DC.

Eswaran, H., Lal, R., and Reich, P.F., 2001. Land Degradation :An Overview. *Proceedings of the second International conference on Land Degradation and Desertification,* Khon Kaen, Thailand, Oxford Press

FAO, 2003a. World agriculture: Towards 2015/30. *An FAO perspective,* edited by J. Bruisnsma, FAO, Rome and London, Earthscan.

FAO, 2006b. FAO statistical databases. Rome

Galloway, J.N., Dentener, F.J., Caponi, D.G., Boyer, E.W., Howarth, R.W., Seitznger, S.P., Asner, G.P., Cleveland, C.C., Green, P.A., Holland, E.A., Karl, D.M., Michael, A.F., Porter, J.H., Townsend, A. R., Vorosmarty, C.J. 2004. Nitrogen cycles: past, present and future. *Biogeochemistry,* 70: 153 – 226.

Gleick, P.H., 2000. Water futures: A review of global water resources projections. World water scenario: Analyses. *Earthscan Publications*, London, pp. 27-45.

Inforesources – Focus. 2006. *Water for food. Focus no. 3/06.*

IPCC 2001b. Climate Change 2001: The Scientific Basis. J.T. Houghton, Y. Ding, D.J. Griggs, M. Noguer, P.J. Van der Linden, X. Dai, K. Maskell, and C.A. *Johnson eds. Cambridge*, UK. Cambridge University Press:881.

Jodha N. S. 1992. Common Property Resources and Dynamics of Rural Poverty in India's Dry Regions. FAI (Fertilizer Association of India) , 2002. *Fertilizer statistics: The fertilizer association of India*, New delhi.

Jones Lang La Salle, 2005. India the next IT offshoring locations. Tier II cities.

King, B.S., Tietyen, J.L., and Vickner, S.S., 2000. Consumer trends and oppurtunities. Lexington, USA, University of Kentucky.

Kumar, P and Birthal, P.S., 2004. Changes in consumption and demand in livestock and poultry products in India. *Indian Journal of agricultural marketing*, 18(3): 110-123.

NOAA,2006. Trends in atmospheric Carbon dioxide. *NOAA/ Earth System Research Laboratory*, Global Monitoring Division.

Oenema, O., Oudendag, A and Velthof, G.L., 2007. Nutrient losses from manure management in the EU. *Livestock science*, 112: 261 – 272.

Rabalais, N.N., 2004. Eutrophication in the Sea. *The global coastal ocean: Multiscale Interdiciplinary processes* Vol 13 Pp 819 – 865. eds. A.R. Robinson and K Brink. Harvard. Harvard University press.

Rae, A 1998. The effects of expenditure growth and urbanization on food consumption in East Asia: a note on animal products, *Agricultural Economics*, 18(3): 291-299.

Roost, N., Molden, D., Zhu, Z., and Loeve, R., 2003. Identifying water saving oppurtunities: examples from three irrigation districts in China's Yellow river and Yangtze River basins. *Paper presented at the 1st International Yellow river forum on River basin Management held in Zhengzhou*, China, 12 – 15 May 2003.

Schmidhuber, J and Shetly, P, 2005. The nutrition transition to 2030. Why developing countries are likely to bear the major burden. *Acta Agriculturae Scandinavia.* Section C. Economy, 2 (3-4): 150-166.

Shankar, V. and Gupta, J. N., 1992. *Restoration of degraded Range lands*. In : (Ed) Singh, J.S., Restoration of degraded range lands-Concepts and strategies. *Rastogi Publication*, Meerut, India. Pp 115 – 165.

Singh, O.P., Sharma, A., Singh,R., Shaw, T. 2004. Virtual water trade in dairy economy. *Economic and political weekly*.

Soto, Ana M., Jannie M. Calabro, Nancy V. Prechtl, alice Y. Yau, Edward F. Orlando, Andreas Daxenberger, Alan S. Kolok, Louis J. Guillette, Iris G Lange, Caros Sonnenschein.2004. "Andrgenic and estrogenic activity in water bodies receiving cattle feedlot effluent in Eastern Nebraska, USA". *Environmental Health perspective*. Vol.112 (3) pp 346-352. March 2004.

UCA (Union of Concerned Scientists) "Chapter 1: Antimicrobial Resistance", in Hogging it! : *Estimates of antimicrobial abuse in Livestock*. UCS January 2001:5.

UN, 2005. World population prospects. The 2004 revision. *UN Department of Economics and social affairs*. NY, USA.

W R I (World Resources Institute), 2005. *Earth Trends: The Environmental Information Portal*

William C Meehan and William S Platts, 1978. *Journal of soil and water conservation*, Nov – Dec 1978. Vol. 23, no – 6.

* * *

17

STUDIES ON PRODUCTION PERFORMANCE OF DAIRY ANIMALS USING PADDY STRAW BASED COMPLETE FEED

A.K. Mishra[1], J.B. Singh[2] and A.K.Gaur[3]

[1] Assistant Professor, Jawaharlal Nehru Krishi Vishwa Vidyalaya, College of Agriculture, Tikamgarh (M.P.) 472001
E-mail: gkripa999 @ rediff mail .com

[2] Senior lecturer, MGCGV, Chitrakoot, Satna (M.P.)

[3] AssociateProfessor, Department of LPM, Veterinary College Jabalpur (M.P.) 482001

ABSTRACT

An experiment was conducted in lactating cows and buffaloes comprising of three groups having buffaloes and cows in each groups. These animals were maintained on conventional feeding system with green roughage (Group I) while Group II were maintained on dry roughage as chaffed paddy straw and concentrate fed separately and Group III were maintained on dry roughage as chaffed dry paddy straw soaked in water overnight and blended with concentrate fortified with minerals and vitamins as complete feed /sani. The average daily feed intake (kg) and dry matter intake (kg) of buffaloes in Group I, II and III were estimated as 14.74 ± 2.08, 14.52 ± 1.32 and 16.72 ± 0.89; 13.41 ± 0.07, 13.21 ± 0.86 and 11.32 ± 1.04, respectively. While the average daily feed intake (kg) and dry matter intake (kg) of cows were estimated as 11.17 ± 1.26, 11.77 ± 1.56 and 12.01 ± 1.03; 10.16 ± 0.97, 10.71 ± 1.26 and 9.80 ± 1.12, respectively. The average daily feed intake (kg) and dry matter intake (kg) of buffaloes and cows in Group I and II were more or less similar but significantly lower in feed intake ($P<0.05$) and significantly higher in dry matter intake ($P<0.05$)than those of Group III. The average daily milk yield, fat, solid not fat and total solids in experimental animals indicated a significantly higher milk yield ($P<0.05$) in Group III in both cows and buffaloes. Results indicated that the animals maintained in Group III performed better as compared to other groups.

Keywords: Buffalo, cow, feeding practice, feed intake, milk.

Introduction

Feeding system in our country varies from place to place, availability of feed stuff and purpose of rearing. The dairy animals that are kept in urban area are provided concentrate and roughages in which concentrate is well supplemented with minerals and vitamin, and roughages are mainly straws with due weightage of the requirement. In peri urban areas animals are provided with green and dry roughages and small quantity of concentrate. While in rural area dairy animals are fed roughages and concentrate as per the availability in the area. Since country is in short supply of feed and fodder, there is need to explore new feed resources and conventional feeds which can be fed to the animals Pandey (1995). There is scope and need for improvement of poor qualities crop residues by different processing technologies Waje (1996). The feeding of milch cows and buffaloes with complete feeds based on concentrates, minerals and vitamin supplements blended with available crop residues like paddy straw will improve the overall performance and health of animals. The 'sani' system followed by most of the dairies suffered from the problem of imbalance of nutrients in a ration as they do not care to supply the essential nutrients such as calcium, phosphorous, vitamins and protein in balanced way. It is, therefore, necessary to balance 'sani' system in respect of these nutrients; such ration will form a complete feed. The roughage and concentrate are fed separately or in the form of sani; looking to these variations in feeding practices, present study was planned to find out the effect of different feeding practices on feed intake, dry matter intake, milk yield and milk composition in cows and buffaloes.

Materials and Methods

An experiment on feeding trial was conducted at Livestock farm, Adhartal, Jabalpur (M.P.) in lactating animals in March, 2003 for a period of 90 days. Experiment comprising of three groups; having lactating three buffaloes and three cows (6 animals) in each group. They were distributed randomly as per parity and milk production records. The proximate composition of experimental feed (D M basis) is presented in Table 1. the animals of Group I were maintained on conventional feeding system with green roughage while Group II were maintained on dry roughage as chaffed dry paddy straw and concentrate fed separately and group III were maintained on chaffed paddy straw soaked in water overnight and blended with concentrate fortified with minerals and vitamin as complete feed / sani. The daily feed and dry matter intake (kg), milk yield (lit.) and composition of milk (%) were recorded in three groups. The analysis of data was done as per Snedecor and Cochran 1989.

Table 1: Proximate composition of experimental feed (% on D M basis)

Feed ingredient	DM	CP	EE	CF	NFE	Ash
Conc. Mixture	91.00	15.66	3.38	14.98	57.96	7.92
Coplte feed 'sani'	65.00	13.30	3.95	23.41	46.36	13.08
Paddy straw	90.5	2.27	1.05	35.50	44.90	16.80

DM: Dry matter, CP: Crude protein, EE: Ether extract, CF: Crude fiber

NFE: Nitrogen free extract

Result and Discussions:

1. Proximate composition of experimental feeds: Crude proteins of concentrate mixture and complete feed sani were estimated as 15.66 and 13.30 percent respectively. (Table 1). The DM, CP, EE, CF, NFE and Ash of concentrate mixture were estimated as 90.0, 15.66 3.38, 14.98, 57.96 and 7.92 respectively while it the corresponding figures of these nutrients were found as 65.0,13.30, 3.95, 23.41, 46.31 and 13.08 percent respectively in group III, where animals were fed feed and fodder together mixed in the form of sani. The values obtained from proximate composition of experimental feeds are in close agreements with Reddy *et al.,* (1996). The proximate composition of paddy straw was calculated and its DM, CP, EE, CF, NFE and Ash were found as 90.50, 2.25, 1.05, 35.56, 44.50 and 16.80 percent respectively. The crude fibre value of paddy straw is more or less similar to the values of Das and Ghosh (2007).

2. Roughage concentrate ratio: The roughage concentrate ratio was 45:55 in groups I and II, whereas roughage and concentrate ratio of complete feed sani was 40: 60 in group III (Table 2). Senger *et al.* (1999) opined that ratio of 45:55 between roughage and concentrate was found satisfactory.

3. Daily feed and dry matter intake: The average daily feed intake (kg) of buffaloes was observed 14.74 ± 2.08, 14.52 ±1.32 and 16.72 ± 0.89 for group I, II and III respectively whereas daily dry matter intake of these groups were calculated as 13.41±0.07, 13.21± 0.86 and 11.32±1.04 respectively (Table 3). The average daily dry matter intakes (kg) of cows was observed 11.17±1.26, 11.77±1.56 and 12.01±1.03 while in buffalo it was 10.16±0.97, 10.71±1.26 and 9.80±1.12 for group I, II, and III respectively. The present findings are in close agreement with findings of Agnew (1996) and Misra, A.K. (1995). The feed intake of cows and buffaloes of group I and II were more or less similar, but significantly lower ($P<0.05$) than those of group III, but the average daily dry mater intakes in cows and buffaloes for treatments I and II are almost similar but significantly higher ($P<0.05$) than those of group III.

4. Milk yield and composition: The average daily milk yield (litre) of cows of group I, II and III were observed as 7.38 ± 0.90, 7.80 ± 0.86 and

8 .68 ± 0.41 litres respectively while corresponding figures of buffaloes of group I, II and III were estimated as 5.90 ± 0.59, 5.95 ± 0.86 and 7.18 ± 0.81 litres, respectively, (Table 4) indicating significantly higher milk yield (P<0.05) in animals of group III which were fed in the form of sani, in cows and buffaloes both. The findings are similar to opinion of Muller and Botha (1998). The average fat percent in milk of the cows of groups I, II and III were estimated as 4.6 ± 0.12, 3.94 ± 0.17 and 3.74 ± 0.59 respectively while the corresponding values of fat content in buffalo milk were 6.68 ± 0.06, 6.84 ± 0.97 and 6.86 ± 0.04 respectively. The impact of dry matter intakes per litre milk production among the treatment groups was close in agreement with Iketaki et al (1996). The SNF percent in the milk of cows fed group I, II and III were 8.56 ± 0.31, 8.66 ± 0.76 and 8.73 ± 0.97 whereas in the buffaloes SNF percent were observed as 9.71 ± 0.08, 9.64 + 0.07 and 9.93 ± 0.57 in group I, II and III respectively. The values did not differ significantly among the different treatment groups of animals. The total solid percent in the milk of cows was 12.71 ± 0.03, 12.67 ± 0.03, 12.81 ± 0.40 in group I II and III respectively similarly in buffaloes milk T.S. percent were 16.34 ± 0.67, 16.54 ± 0.17 and 16.76 ± 0.71 in group I, II and III respectively, showing no significant difference in total solid in milk of cow and buffaloes between treatment groups.

Table 2: Roughage concentrate ratio for the experimental groups

Groups	**Roughage : Concentrate**
I	45: 55
II	45:55
III	40:60

Table 3: Average daily feed and dry matter intake of experimental animals

Groups	**Cows**		**Buffalo**	
	Feed Intakes (Kg)	**DM Intakes (Kg)**	**Feed Intakes (Kg)**	**DM Intakes (Kg)**
I	11.17b ±1.26	10.16a ± 0.97	14.74b ±2.08	13.41a ±0.79
II	11.77 ab ±1.56	10.71a ±1.26	14.52b ±1.32	13.21a ±0.86
III	12.01a ±1.03	9.80b ±1.12	16.72a ±0.89	11.32b ±1.04

5. Digestibility of nutrients: Digestibility of dry matter in the group I, II and III was 54.26 ± 0.36, 48.94 ± 0.76 and 59.58 ± 0.60 respectively for cows while dry matter digestibility among buffaloes were 58.20 ± 0.61, 50.16 + 0.64 and 62.28 ± 0.76, respectively (Table 5). The significant (P<0.05) difference were observed among the treatment groups however in group III significantly (P<0.05) higher digestibility of dry matter in cows and buffaloes. The digestibility coefficient of crude protein in the group I, II and III was 56.75 ± 0.83, 47.95 ± 0.56 and 57.56 ± 0.58 respectively cows while, CP digestibility is buffaloes were 58.90 ± 0.50, 50.55 ± 0.15 and 58.99 ± 0.58 respectively. There was significantly (P<0.05) lower CP digestibility, in which roughage and concentrate were consumed separately. The digestibility coefficient of ether extract for group I, II and III was 62.49 ± 1.0, 44.33 ± 0.90 and 63. 49 ± 0.16 respectively in cows while, in buffaloes it was 67.47 ± 0.20, 48.31 ± 1.0 and 67.24 ± 0.35 respectively. In all three groups difference were found significant (P<0.05) among themselves, group II showing significantly (P<0.05) lower EE digestibility coefficient in cows and buffaloes both. Crude fibre digestibility was 61.17 ± 0.22, 55.38 ± 0.62 and 63.25 ± 0.30 respectively in cows whereas in buffaloes 65.73 ± 0.32, 58.49 ± 0.22 and 66.31 ± 0.63 respectively, in group I, II and III. There was significantly (P<0.05) higher crude fibre digestibility was observed in cows and buffaloes of group III. Digestibility of nitrogen free extract in group I, II and III was 54.64 ± 0.67, 53.54 ± 0.13 and 57.86 ± 1.1 respectively in cows whereas in buffaloes it was 55.09 ± 1.1, 59.17 ± 1.0 and 61.92 ± 1.7 respectively. Higher NFE and ash digestibility on complete feed was in accordance with Yadav et al. (2001). Ash content digestibility for group I, II and III was 30.12 ± 0.84 , 30.25 ± 0.80 and 27.65 ± 0.13 respectively in cows whereas in buffaloes ash digestibility was 36.62 ± 0.74 , 35.92 ± 0.82 and 38.44 ± 0.28 respectively. The difference in digestibility coefficient of Ash in group I and III was non significant (P<0.05) but group II showed significantly (P<0.05) lower digestibility. Significantly (P<0.05) higher digestibility of dry matter on complete feed (Group III) reported by Nisal (2000).

Table 4: Average daily milk yield and milk composition of experimental animals

Parameter	I		II		III	
	Cow	Buffalo	Cow	Buffalo	Cow	Buffalo
Milk yield (litres)	7.40 ± 0.90	5.93 ± 0.59	7.50 ± 0.76	5.99 ± 0.86	8.91 ± 0.41	7.39 ± 0.81
Fat %	4.60 ± 0.12	6.68 ± 0.07	3.94 ± 0.17	6.84 ± 0.97	3.74± 0.59	6.86 ± 0.04
SNF%	8.56 ±. 0.32	9.71 ± 0.08	8.66 ±0. 76	9.64 ± 0.07	8.73 ± 0.97	9.93 ± 0.57
T.S. %	12.71 ±0.03	16.34 ± 0.67	12.67 ± 0.03	16.54 ± 0.17	12.81 ±.0.40	16.76 ± 0.71

Table 5: Average digestibility coefficient of nutrients

Nutrients	I		II		III	
	Cow	Buffalo	Cow	Buffalo	Cow	Buffalo
DM	54.26 b ± 0.36	58.20 b ± 0.61	48.94 c ± 0.76	50.16 c ± 0.64	59.58 a ± 0.60	62.28 a ± 0.76
CP	56.75 a ± 0.83	58.90 a ± 0.50	47.95 b ± 0.56	50.55 b ± 0.15	57.56 a ± 0.50	58.99 a ± 0.58
EE	62.49 a ± 1.0	67.47 a ± 0.20	44.33 b ± 0.90	48.32 b ± 1.0	63.49 a ± 0.16	67.24 a ± 0.35
CF	61.17 b ± 0.22	65.73 a ± 0.32	55.38 c ± 0.62	58.49 b ± 0.22	63.25 a ± 0.30	66.31 a ±0.63
NFE	54.64 ab ± 0.67	50.09 b ± 1.1	53.54 ab ± 0.13	59.17 b ± 1.0	57.86 a ± 1.1	61.92 a ± 1.7
Ash	30.12 a ± 0.84	36.62 a ± 0 .74	30.25 a ± 0.80	35.92 a ±0.82	27.56 b ± 0.13	38.44 a ±0.28

Summary and Conclusions: The following conclusions can be drawn from the study:-

1. There was significantly higher feed intake and dry matter intake in those groups which were fed complete fed in the farm of sani (Group III) than group II and control group I.

2. Daily milk yield was highest in those animals which were fed complete feed sani supplemented with mineral mixture and lowest in control group I and II.
3. There was significantly higher nutrient digestibility coefficients (DM, CP, EE, NFE and Ash) were concluded in cows and buffaloes of group III. Poorest digestibility was shown by the animals of group II and Medium digestibility of nutrient was observed in group I.

The present study indicated that significantly higher dry mater intake, least cost of feed and cost of milk production per liter of milk/day and also significantly highest daily net income per day was recorded in the animals which were fed with complete feed in the form of sani made by soaked paddy straw and concentrate mixture than control group. As it is known that the ultimate goal of modern dairy farm is to increase productivity per unit of cost so as to achieve high returns. So it seems from present study that feeding of paddy straw based complete feed to crossbred cows and buffaloes are more feasible for better performance of milk production and high rate of returns.

References:

Agnew, K.W. 1996. An examination of the effect of method and level of concentrate feeding on milk production in dairy cows offered grass silage based diet. *Indan J. Anim. Sci.* 63 : 21-31.

Das, A. and Ghosh, S.K. 2007., Effect of concentrate supplementation on intake, digestion and milk production in crossbred cows grazing on native pasture of Tripura. *Indian J. Anim. Nutr.,* 16 (3): 238-243.

Iketaki, T., Ota, S., Hamamura, K. and Suzuki, S. 1996. *Research Bulletin of Obihiro University.* 12(3): 201-206.

Misra, A.K. 1995. Effect of restricted feeding of urea treated wheat and rice straws on nutrient intake and digestibility. *Indian J. Anim. Nutr.,* 12 (4): 222-224.

Muller, C.J.C. and Botha, J.C. 1998. The comparative performance of primiparous Holstein Friesian and Jersey cows on complete diets during summer climate. *South African Journal of Animal Science,* 28: 161-166.

Nisal, P.R. 2000. Performance of Holstein Friesian and Jersey heifers on complete feed based on forest grass. Proceedings on IIIrd Biennial A.N.A. conference.

Pandey, U.V. 1995. A Profile : *Indian Journal of Agricultural Economics,* July-Sept. 2003.

Senger, S. S., Tomar, S. K. and Katiyar, R. C. 1999. Feed intake and nutrient utilization in buffalo heifers fed paddy straw based complete feeds. *Indian J. Anim. Nutr.,* 16 (4): 362-365.

Snedecor, G.W. and Cochran, W.G. 1989. Statistical Methods. 6th Edn. Oxford and IBH Publishing Co., Calcutta.

Waje, S.H. 1996. Effect of compounding and pelleting of feed on the performance of cross bred calves. *M.V.Sc. & A.H. Thesis*. J.N.K. V.V., Jabalpur.

Yadav, K.K., Lohan, O.P. and Berwal, R.S. 1991. Dry matter intake and digestibility of Basmati and coarse variety paddy straw as compared to wheat straw. Indian J. Anim. Nutr. 18 : 156-159.

* * *

18

POULTRY FARMING AS A MEANS OF LIVELIHOOD FOR SMALL FARMERS OF GARO HILLS

B.K.Mishra
Department of RDAP, North-Eastern Hill University,
Tura campus, Tura-794002
E-mail: drbkm1972@yahoo.co.in

ABSTRACT

As in many other developing countries, poultry farming is a very important component of livelihoods in Garo Hills of Meghalaya, especially in rural communities. In Garo Hills, however, poultry farming does not receive the attention as it deserves and requires, probably because of a lack of understanding about the role that poultry plays in poor households. In many local societies poultry activities, especially the rearing of chickens, ducks, turkeys and, recently, guinea fowls, have traditionally been part of rural households' coping strategies for emergencies that require the quick conversion of physical capital into social, financial and cultural capital. The lack of attention to the poultry sector results in weak veterinary/extension services, and weak and outdated laws and policies. Weak veterinary services increase livelihood uncertainty among poor farming households and compound the risks of disease outbreaks, as disease control/management systems are lacking. This poses a major obstacle to the success of any centrally orchestrated disease campaigns or identification.

In the state poultry farming is still in its infancy, especially in rural areas where traditions are still supreme, and there is need for greater understanding of the role poultry plays in people's livelihoods. Disease control policies often pay insufficient attention to the people whose livelihoods are affected by disease outbreaks. This raises the question as to whether or not control and prevention strategies can work without the participation of those whose livelihoods depend on poultry. There is need to develop understanding about how small land holder producers and village communities are involved in the control and prevention of some of the devastating disease outbreaks that affect people's livelihoods. Key to any control and prevention scheme is an understanding of market access within prevailing poultry production systems, along with holding sizes, level of intensification, overall densities and geographical/ecological settings. This requires a full exploration of how excessive movement of poultry products can either create problems for or enhance local livelihood patterns, through studying how poultry activities are carried out in different parts of the country: How does an outbreak of disease, such as Influenza, create a deficit in the supply and demand chain? How does this affect the livelihoods of poor rural households and communities? Are government efforts in place to cater for the uncertainties of poor rural households? State interventions to control and prevent HPAI aim to control outbreaks through short-, medium- and long-term measures. This strategy emphasizes the need to strengthen veterinary infrastructure within state, strengthen laboratory diagnostic resources, enhance surveillance for outbreaks of Influenza and other trans- boundary animal diseases, develop response capacity, and ensure resources for effective response. However, this focus on control and treatment outcomes emphasizes the biomedical model, but ignores show households and communities adjust to the loss of poultry in their livelihoods.

Introduction:

The term 'Poultry' collectively reveals all the birds which have been domesticated for their eggs, flesh or feather, for fighting or for pets, and includes chickens, turkeys, geese, guineas, pigeons, swans and peacocks. The fowl is used to refer to poultry in general, and is used in reference to domestic chicken both hens and cocks.

Poultry production particularly in Animal Husbandry plays a very important role for the production of protective food in human nutrition. It carter as a source of supplying vital nutrients like essential amino acids, minerals and all needed vitamins. Poultry keeping is also an important industry in our country in which millions of people are at present engaged for their livelihood.

It is believed that the modern fowl probably originated from following four breeds viz. the Red Jungle fowl (Gallus gallus), the Ceylon Tangle fowl (G. lafayetti), and the Grey jungle fowl (G. sonnerattii) and the Java Jungle fowl (G. varius). It is likely that of the four wild species Gallus gallus is the main ancestor. The four species are closely related and are known to interbreed. India and throughout mainland of Southeast Asia to the offshore islands of Indonesia are the ancestral home of the present day domestic fowl.

Domestication of fowl started in Iran by 800 B.C and the Persians spread throughout Western Asia and to the shores of the Mediterranean by 600 B. C. They were common in Italy by 400 B.C and were introduced into northern Europe by 100 B.C. Chickens were imported into the America and Australia by the early Europeans explorers and immigrants. To-day they are considered as a vital source of animal proteins throughout the world.

In India our history shows that the people were quite familiar with fowl, duck, quail, and turkey by about 3,200 B.C. But the chicks were considered neither a food nor egg machine but something of an entertainer. Quail and cock fight provided as a source of fun. Scientific poultry keeping was first advocated by Christina missionaries towards the end of the 20th century A.D. Their flocks of exotic breeds excelled in performance and were far superior to those of the deshi fowls. The first Mission Poultry Farm was established in Eta, U. P. in 1912 and the first poultry exhibition was held at Lucknow in December of the same year. It was only in 1957, under the Second Five-Year Plan that an organized effort was started under All India Poultry Development Project and regional poultry farms were set up in Bangalore, Bombay, Bhubaneswar, Delhi and Shimla to acclimatized imported good quality stock, propagate and provide training facilities. Chicks of White Leghorn and Rhode Island Red breeds were imported from U.S.A. to be used in these farms besides 269 poultry extension-cum-development centres were established. Since then India has recorded a tremendous growth in the last five decades and specially in the last two decades it has made a spectacular progress and is evolving into a full-fledge commercial, organised and scientifically based agro-business providing employment to thousand of poultry farmers, Besides a network of more than 500 hatcheries have also been established throughout the country, which has resulted in the formation of large, medium and small farms all over the country.

Poultry industry has also achieved satisfactory measures of self-sufficiency in the manufacture of compound poultry feed equipments and machinery, pharmaceutical and biological etc. Including human resource development (HRD) and research. The Indian Council of Agriculture (ICAR) is playing a vital role for poultry development in the country. It has established two poultry institutes which are-

1. Central Avian Research Institute at Izzatnagar (CARI), Bareilly, U.P.
2. Project Director on Poultry (PDP) at Hyderabad, A.P.

To provide necessary research training and extension support to the growing Indian Poultry Industry. There are about 60,000 farms and 100 Intensive Production Centres utilizing advance technology with goods developed either locally or imported from other countries. Some of these farms are very big having capacity of more than one lakh birds. Besides NABARD is also playing an important role in providing financial support to farmers and unemployed youths from day to day due to the impact of modernization and industrialization though this is still far for people of rural and tribal areas of the country. In these areas the people mostly follow backyard open range system of poultry keeping with flock ranging from 25-250 birds. The indigenous native fowl and their cross still continue to be the main stay of these rural poultry farms.

At present in India the per capita availability of eggs and meat is 34 and 700 gm per year, respectively, as against the recommendation of 180 eggs and 11 kg meat annually for a balanced diet. Since poultry meat is cheap and preferred by consumers, it occupies 80% of the market and thus represents 9 kg per capita annually. India rank 5th and 22nd in the world in terms of egg and poultry meat annually. The production comes from about 90,000 farms mainly located in urban and peri- urban areas. To fill the present, gap an additional production of 150 billion eggs and 8.44 billion broiler birds per year is required.

Andhra Pradesh is the leading stated in poultry production and hence is considered to be the poultry capital of the country. More than 60% of egg produce comes in India from following five states viz. Andhra Pradesh, Maharashtra, Punjab, Tamil Nadu and Haryana. Similarly, more than 60% of the poultry meat is produce in Andhra Pradesh, Maharashtra, Punjab, West Bengal and Karnataka. About 75% of the table egg is directly cooked while all the produce in the country is consumed by 25% of the population.

The poultry industry in India is primarily chick oriented and chicken accounts for more than 90% of the total poultry available in the country. Duck next to chicken in order of preference accounts for about 6% of poultry production.

Poultry Farming in Garo Hills: State Profile: Meghalaya

Meghalaya is one of the North Eastern states of India. Meghalaya is bordered on the south and southwest by Bangladesh and on all other sides by the state of Assam. The area is 22,429 square kilometres. The capital is the hill town of Shillong. One of the wettest region on earth, Cherrapunji lies 56 km from Shillong; which has an average rainfall of 450 inches (11,430 mm) over a 74-year period, the highest ever recorded in Asia and the second highest in the world. The state has seven administrative districts-the East and West Garo Hills, the East and West Khasi Hills, Jaintia Hills, Ribhoi and South Garo Hills.

Garo Hill District:

The Garo Hills district was divided into two districts, viz. The West Garo Hills and East Garo Hills districts in October 1976. The West Garo Hill district was further divided into two administrative districts of West and South Garo Hills is June 1992. The district headquarters of South Garo Hills is Baghmara. As per 2001 census the population of the 3 district is 7, 26,800. The people of the districts depend primarily on Agriculture and Craft sectors for their livelihood. The main crops grown are paddy, mustard, jute, cotton, potato, cashew nut, areca nut, pineapple etc. Wet rice cultivation is prevalent in the hills. A sizable number of villagers depend on handloom and cane crafts for their livelihood. There are more than 50 cooperatives involved in Handloom and Textile activities.

Administrative details:

GEOGRAPHICAL Area (Sq Km)	7464
No of Blocks	17
No of Villages	2726

The People:

The people of Garo Hills are who called themselves as Achik mainly depends on agriculture for their livelihood. Like any other tribal Jhum cultivation is the primary system of cultivating the land. Though the land is rich in minerals, forest and other resources the people were very poor but after independence, various schemes have been adopted by the central and local governments for economic development. The district council was created under the sixth scheduled of the Constitution of India to preserve the traditional way of life of the people, to protect them from exploitation by others, and to make them leaders of their own progress. The tribal development blocks came into existence to ensure speedy development. Incentives provided for cash crops and efforts made to popularize terrace cultivation have yielded some results. The communication bottle-neck, soil erosion and loss of fertility are however, among the problems retarding prosperous economic growth. The occupational mobility is a noticeable phenomenon. The literacy rate and educational status is increasing and the persons are also involving in professions other than traditional agriculture. Many are employed outside Garo Hills. Some are wealthy class. The growth population and markets inspired many to set themselves in business. Poultry, dairy farming and bee-keeping are also practiced.

Poultry Farming:

Since time immemorial backyard poultry farming has played an important role to meet the domestic as well as socio-cultural needs of the rural people. The Garos also practice backyard system of poultry farming since long but in recent years the scenario has been changing due to the efforts of the governments and several Non Government Organizations present in the area. Several privately own small and medium farms have come up in semi-urban areas practicing the 'all-in-all-out' system of rearing broilers. Besides these government has set up three poultry farms in the district at Rongram, Simsangiri, Baghmara and Phulbari for supplying chicks, eggs and broilers. These farms are allowed to maintain breeding stock, table and hatching eggs. This farm also aims at distributing exotic breeds and providing extension support to farmers. The per capita availability of poultry meat and eggs is 448 gm and 47 nos. respectively per annum against recommended norms of 1000 gm meat and 180 eggs. Therefore, there is ample scope for development in this sector.

Performance of Government poultry farms in Garo Hills:

Sl. No.	Name of the Farm	Particulars of Birds reared		Production/Supplied		
		Chicks/ Growers	Layers	Eggs Production '000' No.	Birds Layers/ Growers sold for Breeding etc.	Broiler reared and Sold
1	Poultry farm; Rongkhon	2218	294	36.7	1372	6041
2	Poultry farm; Simsangiri	920	460	79.1	425	1377
3	Poultry farm;Baghmara	-	292	21.2	-	-
4	Poultry farm; Phulbari.	-	-	-	-	-

Why Poultry Farming?

Poultry farming can play a vital role in providing employment to the unemployment to the unemployed youth and farmers because it has the following benefits-

1. **It requires minimum investment to start:** In comparison to other livestock it requires less investment. Persons form low income groups may also groups may also starts the business as backyard venture.
2. **Rapid reform of profit:** Chicken starts laying eggs when they are about 6 months of age. Broilers get ready for market at 6-8 weeks of age.
3. **Poultry convert food protein efficiently:** Among all domesticated animals, a broiler takes only 1.9 kg feed for 1 kg foods protein.
4. **Poultry provides continuous source of income:** Since they start laying at their age of 6 months the farmers also start getting return so early, broilers ready for market within two months.
5. **Farming requires small space with confinement rearing:** Poultry can be produce in the backyard of the cities and small towns.

6. **It stabilizes farm income:** farmers occasionally experience group failure due to unfavourable weather condition when poultry rising as mixed farming will increase to stabilize farm income.
7. **Poultry feeds not commonly used for human beings:** Large quantity of Agro-industrial by products which cannot be used for human food and it can be used for poultry feeds.
8. **Availability of superior stocks:** Some of the best breeding stock available anywhere in the world is now being multiplied in this country. Numbers of egg producing breeds, meat producing breeds, their crossing are also available in government farms.
9. **Employment opportunity:** Poultry farming offers opportunity for full-time or part time requirements for women, children or elderly people on the farm operation. Apart from direct employment on the farm, person away from the producing centre are also employed for transportation and distribution among the consumers factories for producing liquid eggs. Egg powdered, dried yolk, etc may help for further opportunity.
10. Poultry manures used as fertilizers: Poultry manures are an extremely rich source of nitrogen and organic manures. It contains 1-1.8 Nitrogen, 1.4-1.8 Phosphorus and 0.8-0.9 Potash.

Potential of Poultry Farming:

The poultry farming which was only a mode of living in the primitive stage and was fun and managed only at a small level for domestic consumption alone has been developed as a full-fledge industry in present time. The change in the nature of poultry farming has made poultry industry an employment generating industry and a good income earning enterprise easily. It is also a lucrative enterprise for the persons seeking employment in the self organization. No doubt, by keeping only 1000 birds two persons any get full time employment on an even basis through the year.

The employment potentialities of poultry industry may be in the form of direct and indirect labour. The direct labour involves the keeping, feeding and maintaining birds on regular basis, while the indirect labour involves the job of the preparation of shed: transportation of feeds, eggs and birds;

storage of eggs; marketing of products. Thus, the industry involves huge labour requirement and enormous job opportunities. If the same is managed on a scientific line, this any solve the problem of unemployment to some extent.

The poultry industry has also got good marketing potential and has got huge development possibilities. Seeing the labour and employment potentialities involved, the industry has great employment situation present in the region. To go further with the discussion a model scheme is shown below-

Model scheme for a Poultry Unit (with 200 chicks):

The climatic condition of Garo Hills is suitable for broiler farming. Broiler meat is being getting popularity now a day because of unavailability and high price of other meat. Tura is no any taboos for consumption of poultry meat and is accepted from agriculture and their by product availability locally could be efficiently used for poultry feeding. Broiler farming is feasible as:-

(i) The returns form broiler farming may be obtain in a shorter time and it does not required large capital investment and land.

(ii) Broiler is more efficient converter of crops and other agricultural by products into meat with higher geed conversion efficiency.

(iii) Broiler excels in dressing percentage. They yield about 70% of their body weight.

Assumptions:

(i) Size of unit ________________ 200 chicks.

(ii) Size of batch ________________ 200 broilers.

(iii) Frequency of chick purchase ________________ once in every 2 months.

(iv) Number of shed ________________ one.

(v) Rearing period ________________ 8 weeks.

(vi) Cleaning week ________________ 1-2 weeks.

(vii) System of rearing _______________ Deep litter system.

(viii) Mortality rate _______________ 5%

(ix) Feed consumption/bird at the end of 8 weeks@ 68 gm/bird/day _____ 3.8 kg.

(x) Expected live wt of birds at the end of 8 weeks ________ 1.8 kg.

(xi) Feed cost _______________ Rs.13/kg.

(xii) Feed conversion ratio _______________ 2.1:1

(xiii) Floor space requirements _______________ 1 sq feet/bird.

(xiv) Construction for broiler type _______________ Rs.450/sq ft.

(xv) Equipments _______________ Rs. 15/bird.

(xvi) Miscellaneous including veterinary aids litter material, electricity _______ Rs. 5/bird.

Methods of production: All in –all out system.

Equipments:

1. Feederers
2. Waterers with grill.
3. Brooder.
4. Flame gun.
5. Weighing balance.
6. Measuring jar (while weighing)
7. Transport crates.

1. Estimated cost for setting up a farm for 200 birds. (All in-out):

(A) Capital Cost- (Rs)

(i) Broiler shed for 200 birds = 200.450

= Rs.90, 000.

(ii) Equipment like brooder, feeders, waterers and electricity installations @ Rs. 15/ bird.

= 200.15

= Rs.3, 000.

Total Capital Cost = Rs.93, 000

(B) Recurring cost (for I batch i.e. 8 weeks)

(i) Chick cost @ Rs. 18/chick. = 200.18
= Rs.36, 000.

(ii) Feed cost = Rs.3.8×200×13
= Rs.9880.

(iii) Miscellaneous expenditure @ Rs.5/bird = Rs.200x5
= Rs.1000.

(iv) Labour cost@ 2100/labour for 2 months = Rs.4200.
Total Recurring Cost = Rs.18, 680.

(v) Therefore Total Cost of the Project
= Rs.93,000 +18,680.
= Rs. 1, 11,680.

(C) Income from I batch
(i) Sale of broilers = Rs.190×60×1.8
= Rs.20, 520

(ii) Sale of litter materials (L.S) = Rs.500

Role of the Govt. in Encouraging Self-Employment in Poultry Industry:

India lives in villages and at present the Government is laying an added stress on the development of villagers with rural reconstruction. Poverty Elimination and Huge Employment Generation have been the main objective of all the programs launched by the Govt. during the plan period. In all the plans an added stress is being laid on higher employment generation and eradication of poverty.

It can be said firmly that if one unemployed graduate or undergraduate is provided with finance to maintain a poultry farm with 1,000 poultry birds or so, he can very easily earn his livelihood. He can maintain his family also. Poultry would provide employment to a person throughout the year on an even basis, whereas agriculture is itself a case of disguised unemployment. My stress is that with a view to generate enormous employment opportunities and removing the incidence of poverty amongst the rural folk poultry enterprise should be included in the list of ' top priority items' and a definite percentage of credit be earmarked for this enterprise.

For developing rural India more and more stress is laid on the development of poultry farming with dairying and other rural industries on scientific lines for ensuring higher employment generation and adequate earning to the rural poor. Poultry industry by nature has got both the potentialities. No enterprise of rural economy may be kept at par with poultry and increasing pressure of population on agriculture. Hence the Govt. would have to develop poultry farming effectively and would have to earmark a definite percentage of Bank credit to it by including the enterprise in the list of priority enterprises for providing a solid bade to the enterprise. In developing poultry industry not only in the Region but in the country as whole the role of the Govt. seems to be of immense value.

Conclusion:

Poultry are inseparable from mankind whether in urban slums or the rural scenario not to speak of tribal areas and backward harsh environments. In the rural scenario poultry do not need any land, birds are easy to manage with these hardy birds managing themselves most of the time in a set pattern returning faithfully more for the night and regularly laying eggs at home. They are rugged, disease resistant and well adapted to the harsh rural environment. With better nutrition their egg production can also be stepped up substantially. Even stock multiplication is easy the hen brooding its eggs and tending to the young chicks till they can protect themselves. The essential thing that these hardy birds need is a good Night Shelter. With all these attributes poultry farming in the rural environment can be powerful tool for poverty alleviation an social justice. To the rural poor this can be Rainless Harvest with egg production and stock multiplication proceeding unhampered irrespective of rain or drought.

References:

www.cdc.gov.

www.usda.gov

www.backyardchickens.com

www.aeb.org

www.nationalchickencouncil.org.

http://www.poultryegg.org/

http://www.unitedegg.org/

http://en.wikipedia.org/wiki/poultry_farming

19

A REVIEW ON PRESENT STATUS AND PROSPECTUS OF MITHUN (*BOS FRANTALIS*) REARING IN MANIPUR

Dr. Ch. Nanda Kisore Singh
Veterinary Officer, Directorate of Vet. & A.H. Services, Manipur[1]

Miss Ch. Victoria Devi
Research Scholar, Deptt. Of Agri. Extension, BCKV,
Mohanpur, Nadia, W.B.[2]

ABSTRACT

Mithun (Bos frantalis), the domesticated free-range bovine species, is an important component of the livestock production system of North-Eastern hilly region of India. Thus unique bovine species is believed to be domesticated more than 8000 years ago. Mithun is primarily reared as meat animal and highly preferred among the tribal people of North-Eastern region of India. Mithun is also used as a ceremonial animal and plays important role in economical, social and cultural life of the tribal people of North-East. Besides, it is now established that superior quality milk and hide can also be obtained from Mithun. National Research Centre on Mithun was established by ICAR in the year 1988 in the state of Nagaland to conserve, propagate and improve this species for the future use.

Keywords: Mithun, prospects, limitations, Manipur.

Introduction:

Manipur is located between 23.83^0 North to 25.83^0 N Latitude and 93.030 and 94.47^0 E longitude. The average rainfall is about 967.20 mm. The total area of the state is 22.327sq.km. The state has a total of 9 districts. The four valley district i.e. (i) Imphal East, (ii) Imphal West, (iii) Bishnupur and (iv) Thoubal are surrounded by the five hill districts namely, (1) Churachandpur (2) Chandel (3) Senapati (4) Tamenglong and (5) Ukhrul wherein the tribal people are living mostly depending on shifting Jhumming cultivation. Ninety percent of the total area of the state is hills and remaining only 10 percent forms the valley area. The total forest area is 17,418 sq.km. Out of which 1467 sq.km. are reserved forest, 4171 sq.km. are protected forest and 11780 sq.km. are un classed forest. The total population of the state as per the 2001 census is 23, 88,634.

Mithun (Bos frantalis), one of the rare species of livestock, is found confining only in the hilly tract of North Eastern Region, mostly in Manipur, Nagaland, Mizoram and Arunachal Pradesh. The state has 16,660 Mithun (1997 livestock census) scattering in the five hill districts. The animal has an important place in the social economic and religious life of the tribal people of the state. The animal is reared by tribal only in the hilly regions mainly for meat purposes and is considered as a sign of prosperity and superiority of an individual in the tribal society of the state. The animals are sacrificed in the marriage ceremonies, religious function or big feast. The animals are also offered as dowry by the bride's party to the bride groom.

Mithuns are semi-wild in nature and are found in varying ranges of altitudes from 300m to 3000m above mean sea level in the country. They thrive well in the humid climate. They are well adapted to the hilly terrains at varying degree of slopes. Mithuns are sure footed animals in the hilly region. They can climb easily in sloppy, steep terrain unlike other domestic animals such as buffalo, cattle, etc. The animals roam in forest for months together and sometimes, the owner goes to locate them. Mithun cow calves in jungles sometimes and return with calf to the owner. They are found to thrive well on fodder available in the hills in the form of tree leaves, shrubs and grasses. They are known to utilize the coarse fodders which are left unconsumed by other ruminants. Mithuns are not close grazers like cattle but browsing in nature like goat. The peculiar characteristic of this rare animal is its ability of grazing on the steep slopes of hilly areas of the state. Mithun is similar in appearance to Gaur except in size and horn. The size is smaller and the horns are more or less straight unlike gaur. The animal has huge head, deep massive body and sturdy limbs. The forehead is broad and concave. Breadth between the eyes range approx. in between 9-10 inches in adults. The animal has no hump "above the dorsal ridge but the ball has an exuberance of flesh immediately over the shoulder. It has broad ears and small dewlap. The length of ear ranges approx. from 9" to 10" in adults. The legs and tails are shorter than those of cows. The horns are straight more or less and are either black and white. The horns of young ones are smooth and polished but in adults, the horns are rugged and identified at the base. The length of the horns in adults ranges from 14" – 15". The height at withers ranges from approx. 46" to 53" in adult females and males. Body length varies from 51"to 58" in adults. The neck is short and length varies from 20" to 21" in adults.

The colour of adult Mithun is black and white the shank and forehead are white in colour. White colored Mithun are found rarely. Animals with white patches in an around the ridge and belly are also commonly encountered. A newly born calf in generally light golden yellow which soon changes to fawn, then to light brown and ultimately to coffee or jet black in adults.

Mithuns are essentially hilly animals and they rarely come down to lower altitudes except in quest of pasture because they never prefer plain areas. They come out fro grazing in the morning and graze till the evening. During hot hours, extreme cold or cloudy day, they retire in their shelter and seclusion of the forest. They have the habit of visiting salt licks spots where the ground is rich with salts and minerals like other animals. They are shy and timid animals by nature. The average body weight of Mithun ranges from 400 – 500kgs.

The mating time occurs throughout the years and calves are born in all the seasons. The gestation period is similar to that of cow and ranges from 280 – 5 days. Mithun separates from the herd when her calf is born and remains with it for a few days. The herd also remains in vicinity. She rejoins the herd as soon as the offspring is able to accompany her. In spite of the diversity of the animals in terms of adaptability, quality of meat, thriveness on coarse fodder. Grazing on steep hills, semi wild nature and its importance in socio-economic life of the tribal people of the state, the population of Mithun is found decreasing from 18813 to 9908 during the period from 1984 to 2007.

In spite of the wide scope for Mithun rearing and its economic viability, Mithun rearing was not popularized as it could be and it is rather on its declining trend for the following reasons:

1. Lack of proper planning and development programme which hindered production of better germplasm for propagation and improvement of indigenous Mithun.
2. Lack of awareness on the part of the farmers regarding the profitability of Mithun rearing and over emphasizing on Jhumming cultivation which, however, is no more productive and profitable then Mithun rearing.
3. Lack of purchasing powers and capital fund by the poor villagers for purchase of breeding Mithun.
4. Factors of wild carnivorous animals like tiger, wolves etc. that have killed and eaten up Mithun particularly the young ones in the jungles.
5. Interference with Jhumming cultivation. About 98% of the hill inhabitants are jhumea practicing shifting cultivators. The Mithun habitually living in the jungles are usually having free access to the cultivated land and destroyed or eaten up the crops unless proper fencing was not constructed around the Jhum land. To construct such fencing every year, extra labour are required if Mithun are kept in the village, otherwise not.
6. Massive deforestation programme which has been going on over the years due to population pressure.
7. Indiscriminate slaughtering of the animals.

Population Growth Trend of Mithun since 1984 – 2001:

Sl. No	Name of the District	YEAR						Deviation on the basis of 1984 in%
		1984	1987	1992	1997	2003	2007	
1	Churachandpur	901	924	2070	1283	29288	1345	
2	Senapati	1430	1475	1624	853	99954	1124	
3	Tamenglong	2653	2827	1938	1908	31549	1822	
4	Chandel	1410	1435	3845	2938	36728	4575	
5	Ukhrul	12419	13234	11545	9678	56309	1041	
6	**Total for state**	**18813**	**19895**	**21022**	**16660**	**253828**	**9908**	**(-)47.33%**

The above five hill districts are breeding tracts of Mithun in Manipur. Bases on the population of Mithun as per the 1984 census there has been drastic decline in the Mithun population in the state. The total of 18813 Mithun found in 1984 livestock census were reduced to as low as 9908 in 2007 causing a population decline rate of significant (-) 47.33%. Signaling eminent threat on the existence of this precious meat purpose breed. This threat may turn into reality unless certain immediate significant development programmes for Mithun have not been taken up in the state. The State Government also could not focus on any studies relating to Mithun so far due to paucity of fund.

Conservation of Genetic Resources of Mithun:

1. To conserve and promote the germplasm of semi domesticated Mithun in their natural habitat with the maximum participation of village based farmers for socio economic development of the hill people of Manipur.
2. The tame semi wild natured Mithun by grooming, patting, salt feeding, milking etc. and making it economic animal for production of meat and milk.

3. To produce maximum number of improved breeding stock of Mithun for distribution to interested farmers for rearing.
4. To motivate the farmers for taking up large scale Mithun rearing fro replacement of Jhumming cultivation which is unprofitable and destructive to the forest. It will help in a big way to maintain ecological balance as well.
5. To organize awareness campaigns and impart training to the farmers on economic rearing and management of Mithun.
6. To establish as many as possible Mithun shelter centers in the jungle adjacent to the village where Mithun are on free range condition.
7. To collaborate village based Mithun rearers with the scientific findings and information from national scientist.
8. As the Mithun is one of the best meat purpose breeds, its large scale production/multiplication may be an alternative means to boost the meat production in the country despite banning of cow slaughter. It may take the place of cows in terms of meat production in the country.

Mithun Shelter Centers:

In order to gradually tame the semi wild natured Mithun for domestication, it is considered necessary to establish Mithun shelter sheds in the jungle/forest adjacent to the village where Mithun are available under natural free grazing condition. A suitable hill land sufficient enough for grazing a group of Mithun, say at least 4-6 sq.km area where natural stream and perennial sources of water are readily available be acquired. The shed be constructed with locally available materials at minimum expenditure. Depending upon the situational demand and successful implementation of the scheme more centers may be set up in other ideal places later on.

Feed and Feeding Practices:

Intensive plantation programme of nutritious fodder and edible trees may be taken up for generous feeding of the Mithun. The animals may be encouraged to live on green forages and fodder as they have been doing under natural grazing system. Apart from this, minimum feeding of concentrate @ 1 kg per day per animal with locally available crop residues like, rice polish/bran, wheat bran, broken maize/rice/cakes etc. with common salt and mineral mixture may be practiced, taming and getting quick economic returns from the animals. To facilitate controlled grazing and fodder development, barbed wire fencing with locally available posts be constructed at some strategic selected points with minimum expenditure. The fencing will prevent the animals from interference with other private lands and migration to other distant places.

Breeding Programme:

Natural selective breeding may be practice by maintaining 5:1 female and male ratio. Attempts may be made for taking up cross breeding programme to cross Mithun with cattle by adopting (A.I.) Artificial Insemination.

Strategies for Proliferation of Mithun:

1. Identification of actual numbers of Mithun population and village based Mithun rearer's problems and formation of action plan for their remedial measures.
2. Plantation of nutritious grasses, herbs, shrubs and edible trees for onwards feeding to the animals.
3. Imposition of ban, by proper legislation on killing of breeds able Mithun for meat purpose up to certain age and utilizing such female for breeding and multiplication purpose only.
4. Formation of village based Mithun rearers co-operative society which will be linked up with district and state level Mithun rearers federation.
5. Improvement of marketing facilities and maintenance of uniform selling rate of Mithun and its products.
6. Routine vaccination of Mithun against common disease like F.M.D., H.S. & B.Q. etc. and proper risk coverage under livestock insurance policy.

7. Holding of awareness campaign s for scientific rearing of Mithun and organization of local farmers training and if need be deputation of batches of farmers to National Research Centers of Mithun outside Manipur to acquire more practical knowledge for economic and scientific rearing of Mithun.

Improvement of Traditional Rearing System:

The hill tribals are animal lovers. Traditionally, they reared different types of animals without emphasizing to a particular animal. Mithun was reared in semi domesticated condition in the jungle. They kept them under free grazing conditions in the jungle with zero expenditure on feeding and management. No housing and feeding of concentrate are required because; they can fully depend on natural forges and green vegetation which are abundantly available in jungles. So far, nothing was done for large scale development of Mithun in hill areas of Manipur. If attention is given towards the development of this massive and neglected animal, it can become income generative and alternative to the Jhumming cultivation. Now, the hills tribal who are primarily Jhumming cultivators could not produce crops sufficient for sustenance for their families due to decreased soil fertility following deforestation, erosion of soil and increased population etc. The average annual income of a jhumea is Rs. 10,000(approx.) only. As they could not depend on Jhumming cultivation, they are now looking for an alternative means for earning their livelihood. Meanwhile, many farmers look for animal husbandry like pig rearing, cattle rearing and Mithun rearing for self employment and for earning their livelihood. The cost of grown up Mithun under prevailing rate is Rs. 25,000/- to Rs. 30,000/- only per animal which is more than the average annual income of the jhumea. As such many farmers come to realize the profitability and feasibility of Mithun rearing, and it will be the right time for the government to intervene for the exploration and popularization of Mithun rearing among the inhabitants of Manipur hill areas.

Prospects of Mithun Rearing:

1. Mithun being animal of high altitudes can thrive well naturally in the hills of Manipur with zero expenditure on feeding.
2. Vast area of vegetation, forest land/jungles are available for free grazing of Mithun.
3. All the hill tribals are non vegetarian and there is a high demand of Mithun meat and 1 kg of meat costs Rs.160/-. It is one of the best meat propose breeds.
4. Under traditional free range condition Mithun can attain sexual maturity in 2-2.5 years of age and could calve every year or alternate.
5. Unemployment problem is increasing and many farmers and unemployed youth come to realize the profitability of Mithun rearing for self employment.
6. It holds great promise for food security in view of legislation to ban cow slaughter all over the country.

References:

Reports on 17th and 18th Quinquinal Livestock Census 2003-2007 respectively by Department of Veterinary and Animal Husbandry Service Manipur.

A Handbook of Statistics by Department of Statistics and Economics, 2007, Govt. of Manipur.

* * *

20

LIVESTOCK BY-PRODUCTS AND ITS BIOLOGICAL INDUSTRIAL IMPORTANCE

Anindita D. Sangma
PhD Scholar, Department of Rural Development and Agricultural Production, NEHU Tura Campus, Meghalaya-794001

ABSTRACT

By-products from livestock are an economically important part of the animal agriculture business. By-products yield vary from 55-60% depending on the species of animals. The major by-products from livestock are blood, bones, hoofs, horns, hair, wool, glands, organs, intestines, stomachs, other offal, condemned meat, gut contents, and manure. In this study, the by-products of sheep, swine and cattle are given more emphasis as these are consumed in majority. There are numerous possibilities to use all parts of livestock by-products in the form of keratin, collagen, minerals, fats, fatty acids, natural casings, gelatine, economical alternatives etc., which will help in job creation, income generation, saving of foreign exchange and as a means of removing dirt from the slaughter premises, assuring a sustainable, hygienic and sanitary environment.

Keywords: Livestock, By-products, slaughter, uses

Introduction:

The livestock has a unique place in Indian agriculture. It includes cattle, buffalo, swine, sheep, poultry, honey- bee and fish etc. It contributes to about one fourth of our gross national product. It provides us with milk, draft power, hides, skins, bones, hooves, manures, fuel, meat, egg, wool, honey, wax, fish etc. (Pandey & Singh, 2004)

Livestock production is an important source of income for the rural poor in developing countries, like India, where 70% of the livestock is in the hands of small and marginal farmers and landless labourers, who own less than 30% of the land area. Production Systems are based on low cost agro-by-products as nutritional inputs, using traditional technologies for production. A sizeable percentage of livestock owners are below the poverty line.

Livestock sector is a prominent sector among agriculture and allied activities in India. In 1997-98, livestock alone contributed to 25 percent of total value of agriculture GDP. This has increased gradually from 14 percent in 1980-81. The annual rate of growth in GDP from livestock is 7.3 percent.

Importance of livestock:

Domestic animals are very important to mankind. They furnish precious food products (meat, milk, eggs, honey) and valuable non-food-industrial products (wool, hair, silk, hides, skins, furs, wax, feathers, bones, horns, etc.). Quadrupeds are widely used, particularly in developing countries, as beasts of burden and for draught or are used for commuting to and from agricultural holdings. Some are used also for recreation purposes (horse riding), and most of them are a source of organic/natural fertilizers and fuel.

Feeds of animal origin are also important, e.g. meat meal, bone meal, blood meal, tankage, etc. These are produced from slaughtered animals rejected at the sanitary inspection, from inedible offal, from residues of meat scraps and trimming after the fat has been extracted, from tannery by-products, from poultry by-products (particularly from those processed into ready-to-cook), from hatcheries by-products (infertile eggs and other refuse), from eggshells, etc.

Definition:

The terms "livestock" and "poultry" are used in a very broad sense, covering all domestic animals irrespective of their age and location or the purpose of their breeding. Non-domestic animals are excluded from the terms unless they are kept or raised in captivity, in or outside agricultural holdings, including holdings without land.

Cattle, buffaloes, camels, sheep, goats, pigs, horses, mules, asses and chickens are raised and enumerated in many countries. Some countries raise and enumerate ducks, geese, turkeys and beehives, whereas rabbits, guinea fowl, pigeons, silkworm cocoons, fur animals, reindeer and various kinds of camelids are limited to far fewer countries.

However in simple understanding **"livestock"** means:

(a) Cattle, horses, sheep, swine, poultry and bees,
(b) Animals of a kind raised for meat production or production of products from the animals, including breeding stock of those animals, and
(c) Animals of any species prescribed in the regulations and

"Livestock product" means
(a) Any primary or processed commodity derived from livestock, including meat, raw hides, dressed poultry, eggs, wool, hair and honey, in any form, and
(b) Any part of livestock including offal, fat and by-products

World Meat Consumption by Species(%)

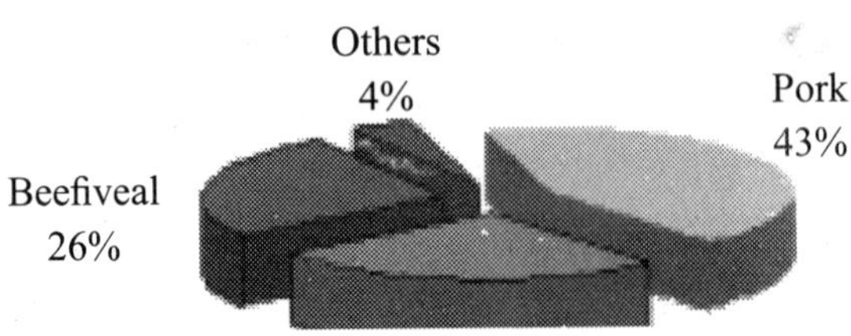

Source: USDA'S FAS

Livestock Population of India:

Last updated: May 17, 2010

Livestock Population in India by Species											
											(In Million Numbers)
Species	**1951**	**1956**	**1961**	**1966**	**1972**	**1977**	**1982**	**1987**	**1992**	**1997**	**2003**
Cattle	155.3	158.7	175.6	176.2	178.3	180.0	192.5	199.7	204.6	198.9	185.2
Adult Female Cattle	54.4	47.3	51.0	51.8	53.4	54.6	59.2	62.1	64.4	64.4	64.5
Buffalo	43.4	44.9	51.2	53.0	57.4	62.0	69.8	76.0	84.2	89.9	97.9
Adult Female Buffalo	21.0	21.7	24.3	25.4	28.6	31.3	32.5	39.1	43.8	46.8	51.0
Total Bovines	198.7	203.6	226.8	229.2	235.7	242.0	262.2	275.7	288.8	288.8	283.1
Sheep	39.1	39.3	40.2	42.4	40.0	41.0	48.8	45.7	50.8	57.5	61.5
Goat	47.2	55.4	60.9	64.6	67.5	75.6	95.3	110.2	115.3	122.7	124.4
Horses and Ponies	1.5	1.5	1.3	1.1	0.9	0.9	0.9	0.8	0.8	0.8	0.8
Camels	0.6	0.8	0.9	1.0	1.1	1.1	1.1	1.0	1.0	0.9	0.6
Pigs	4.4	4.9	5.2	5.0	6.9	7.6	10.1	10.6	12.8	13.3	13.5
Mules	0.1	0.0	0.1	0.1	0.1	0.1	0.1	0.2	0.2	0.2	0.2
Donkeys	1.3	1.1	1.1	1.1	1.0	1.0	1.0	1.0	1.0	0.9	0.7
Yak	NC	NC	0.0	0.0	0.0	0.1	0.1	0.0	0.1	0.1	0.1
Mithun	NA	NA	NA	NA	NA	NA	NA	NA	0.2	0.2	0.3
Total Livestock	292.9	306.6	336.5	344.5	353.2	369.4	419.6	445.2	470.9	485.4	485.0
Poultry *	73.5	94.8	114.2	115.4	138.5	159.2	207.7	275.3	307.1	347.6	489.0
Dogs	NC	NC	NC	NC	NC	NC	18.5	18.0	21.8	25.5	29.0
NC: Not Collected; NA: Not Available * Includes Chicken, ducks, turkey & other birds											
Source: Livestock Census 2003											

From the above latest table of Livestock Census 2003, it was found out that there is a decrease in bovine's population in a decade but there is a decrease of .4 % in the total livestock population. However the population of sheep, goat and pigs has increased.

Major Slaughterhouse By-Products and their Uses:
By-Products

1. **Blood**
   Uses: Liquid blood: used as a source of pharmaceuticals as albumin for the glue, textile and dye industries
   Dried blood: used as blood flour, blood meal for animal feed and as fertilizer.
2. **Bones**
   Uses: Used as animal feed, fertilizer, manufacture of combs, buttons cutlery handlers, glue, gelatin, tallow and ornaments.
3. **Hoofs and Horns**
   Uses: Used as meal, fertilizer, gelatin and glue, manufacture of combs, buttons, hairpins and articles of tourists attraction including souvenirs
4. **Hair and Wool**
   Uses: Used in the manufacture of brushes, yarn, fabrics and fibers
5. **Glands and Organs**
   Uses: Used for pharmaceuticals
6. **Intestines**
   Uses: Used to make sausage casings, musical instruments/strings and surgical ligatures
7. **Stomachs, Other offal, Condemned Meat**
   Uses: Used to make Meat/bone meal, blood, bones, horns, hoofs hair, wool, glands, organs, intestines, stomachs, gut contents, manure for animals feed or fertilizer, tallow for soap and glycerin, lubricants, grease and waxes
8. **Gut Contents, Manure**
   Uses: Used to make compost, biogas as fuel for heating and lighting

Importance of Livestock By-Products:

By-products from livestock are an economically important part of the animal agriculture business. Goldstrand **(1988)** stated that organs, fatty tissues, bones and blood represent **39%** of the live weight of cattle;

economists have placed the value of edible and inedible by-products from cattle at **9** to **12%** of the total live value.

By-products yield vary from 55-60% depending on the species of animals and food value placed on them. Smaller animals have a smaller percent of other by products than large animals.

In less developed countries, the volume of other livestock by products is smaller compared to that in developed countries due to the relatively less number of livestock slaughtered daily and also because most of the by products are used as food. However, in areas where large numbers of livestock are being slaughtered such as in Municipal and City abattoirs, Meat processing plants it is possible to develop the industry.

Development of the by-product industry will help in job creation, income generation, saving of foreign exchange and as a means of removing dirt from the slaughter premises, assuring a hygienic and sanitary environment.

However importance of by- products of some of the important livestock are as follows:

Sheep By-Products:

When most people think of sheep products, two items come to mind: meat and wool. Granted, these two products account for the majority of the income for sheep producers; however, there are numerous possibilities to use all parts of the sheep. In using the carcass to the full extent, everybody, from the producer to the consumer, benefits. The major benefits come from both economical and environmental aspects. By using all parts of the carcass, it becomes more valuable to the processor and as a result, the lamb becomes more valuable, meaning that the producer benefits economically. Most importantly, the consumer saves money in two ways from the processor using the entire lamb. First, the price of lamb meat is reduced because the processor's income increases by using the entire carcass. Second, since sheep byproducts can be sold at a low price, the products that they go into can also be sold at a lower cost. From an environmental standpoint, in using the entire carcass, less waste is accumulated. By producing less waste that needs to be disposed of, everyone benefits from a cleaner environment.

Although by-products provide a great opportunity for everyone in the sheep industry to profit, few in the industry are actually reaping the benefits. It is only through educating those integrally involved in the industry that the full resources of this opportunity may come into fruition.

To begin with, let's divide the parts of a sheep carcass that can be used for by-products and describe how these products can be extracted.

1. From the Intestines.
2. From the Horns, Hooves, & Bone.
3. From Manure.
4. From Fats & Fatty Acids.
5. From Wool.
6. From Hide.

From the Intestines:

In most cases after the intestines are eviscerated, they would be disposed of. However, they can be converted to many valuable products. Once cleaned, the intestines can be used as natural casings for foods such as sausages and frankfurters. This type of casing provides an economical and safe packing material for these popular foods. A high value by-product of the intestines are instrument strings. Though labor intensive, this product can be sold at a high price. Sheep intestines provide a critical product for the veterinary field. Once processed, the intestines are made into suture material. By using natural sutures from sheep, veterinarians avoid complications such as immune responses that arise from synthetic sutures. Also, these sutures are less invasive because the body reabsorbs them over time.

From the Horns, Hooves, & Bone:

The horns, hooves, and bones are rich sources of keratin and collagen among other things. These two proteins are important due to their binding ability and strength. The properties of these proteins allow the horns, hooves and bones of sheep to be used for a wide variety of products. These products include: bone charcoal for high grade steel, shampoo and conditioner, bone china, collagen and bone for plastic surgery, horn and bone handles, bone meal, bone jewelry, adhesive tape, gelatin capsules, bone charcoal pencils, marshmallows, piano keys, gelatin desserts (ice cream, yogart, and jello), film, and shepard crooks.

From Manure:

One of the last places that one would look for a usable by-product is sheep manure. However, due to the ruminant digestive action of sheep, many valuable minerals are found in the manure. The manure contains nitrogen, phosphorous, potassium carbonate, and lower levels of various minerals. Since manure contains these minerals, it is a valuable part of such products as fertilizer and potash.

From Fats & Fatty Acids:

Although ill regarded and of little per unit value, fats and fatty acids can add to the value of the carcass. Due to the high quantities available on the carcass, once fats and fatty acids are further processed, uses can be found for these otherwise worthless attributes of the carcass. Sheep fat and fatty acids can be added to dog and chicken feeds with relatively little further processing. They provide an inexpensive source of energy to fill the needs of the diet. Through further processing, fats and fatty acids can be used as components in wax products. Examples of these products are: paraffin, crayons, candles, and floor wax. Also, fats and fatty acids play an integral role in chemicals; glycol is found in brake fluid and glycerol is the ingredient that makes asphalt stick together. To top things off, most people don't realize that products they use on themselves every day, such as: makeup, cosmetics, tanning lotion, shaving cream, and hand cream, contain sheep fat and fatty acids.

From Wool:

Wool itself is a common product of the sheep industry, and like all other sheep products there is waste material generated that can be converted into profitable items. These products include: lanolin, artists brushes, insulation, and rug pads. Without wool by-products there would be no baseball because: the ball is stuffed with wool, it is sewn with wool thread, its rubber lining is made from stearic acid, and the cork center contains processed blood.

From Hide:

As an alternative to expensive leather from beef hides, sheep hides can be used as an economical alternative. Once removed and processed, the hides can be used in the making of sporting goods such as baseballs, footballs, tennis balls, and basketballs. Also, the leather from sheep hide can replace virtually all leather products made from the leather of beef hide such as footwear and drum heads.

Pork/Swine By-Products:

Pork is the most widely consumed meat in the world. People eat many different pork products, such as bacon, sausage, pork chops and ham. A 250-pound market hog yields about 150 pounds of pork.

Several valuable products or by-products, in addition to meat, come from swine. These include

- insulin for the regulation of diabetes;
- valves for human heart surgery;
- suede for shoes and clothing;
- and gelatine for many food and non-food uses.
- Swine by-products are also important parts of such products as water filters, insulation, rubber, antifreeze, certain plastics, floor waxes, crayons, chalk, adhesives and fertilizer.

Cattle By-Products:

Cattle serve to be one of the amazing animals among livestock with diverse by-product importance. The by-products of cattle are obtained from-

1. wBones and Horns
2. Hide and Hair and
3. Glands and Organs.

   The by-products which are obtained specifically from the above headings are as follows:

Bones and Horns:

- Bone china
- Ice cream
- Piano keys
- Candies
- Knife handles
- Vitamin capsules
- Chewing gum
- Lipstick
- Wallpaper paste
- Combs
- Photo film

Hide and Hair:

- Baseball gloves
- Car upholstery
- Drum heads
- Leather coats
- Violin strings
- Shoes
- Felt hats
- Luggage
- Wallets
- Leather watchbands
- Rawhide softballs

Glands and Organs:

- Asphalt
- Cosmetics
- Fertilizer
- Insulation
- Livestock feed
- Medicines
- Paint
- Plastic
- Soap
- Tires

Apart from the above, some more by- products of importance from cattle are as follows:

Medicinal Properties of cow products: Panchagavya

The medicinal importance of products obtained from the cow knows no bounds. Panchgavya, a mixture of five products of the cow viz. milk, urine, dung, ghee and curd / butter milk is a formulation of Ayurveda, which possesses healing properties against many disorders. Physicians from the Ayurveda system of medicine have been using remedies based on cow's products, either singly or in combination as Panchgavya, for centuries. In recent years, results of scientific investigations in India and abroad into the medicinal and curative properties of these products have created a lot of enthusiasm and hope, as they could be potent anti-cancer and anti-HIV agents.

Cow dung and urine

Cow or cattle dung and urine have an astonishing and myriad variety of uses, which have been developed over a period of time. Some of the important uses for Cow Dung are as follows:

1. Fuel - cow dung patties (goottee) for cooking
2. Fertilizer - composting makes it even more powerful.

3. Heat source - cow dung is naturally hot -compost makes hotter put in glass house to heat glass house or run pipes thru it to get hot water.
4. Purifier - natural antiseptic qualities
5. Floor coating - used mixed with mud and water on floors in mud houses. Improves water absorption of mud. Prevents muddy puddles resulting from spilt water.
6. Mud brick additive - improves resistance to disintegration
7. Skin tonic - mixed with crushed neem leaves smeared on skin - good for boils and heat rash (SP used it for heat rash in Mayapur.)
8. Smoke producer - smoldering cow patties keep away mosquitoes. Can also make smoked paneer over such smoke. Tastes great in pasta!
9. Ash - from patties used in cooking-Pot cleaner - used dry, absorbs oil and fat; wet, as a general cleaner.
10. Brass polisher - tamarind removes oxidation - wet ashes polishes.
11. Fertilizer - alkaline - cow dung ash is basically lime with a few other minerals mixed into it.
12. Mud additive - dries up slippery mud puddles. Mud brick additive - mud and lime (cow dung ashes) becomes like cement
13. Pond PH balancer - thrown into pond neutralizes acid.
14. Tooth polish
15. Sun-dried organic recreational-aerodynamic-device -cow patty Frisbees;
16. Fan for fire - large cow patties can be used as make-shift fans.
17. Deity worship - ingredient in pancha gavya
18. Seed protector- covering seeds in dung before planting helps to protect against pests.
19. Fresh Cow urine taken thirty days straight is an ayurvedic remedy for Brights disease.
20. Disposable camphor lamps for use during fire sacrifices.

21. Another use of cow pies (and camel dung) mix with fresh water till we have a paste and apply to skin diseases. It seems to work to ease the Itching of psoriasis.
22. If you soak your feet in cow urine it will cure athletes' feet.
23. Fresh, less than hours old cow urine seems to have some helpful effect on teen age pimple eruptions.
24. Smoke from Cow-dung or coal actually increases our eyelids to close & open so many times & lot of water from the eyes comes out & the advantage is that it increases the vision life of a person to old age also.
25. For seed-raising.
26. Cow dung use for on insect stings.
27. Cow urine can be used for stones (kidney, gall etc.) shot glass full first thing in the morning fresh from the cow for 21 days (uric acid in cow urine dissolves these stones to a manageable size
28. Apparently, the traditional Indian village system of wiping the floor daily with a mixture of wat30. Urine was used to bleach wool in the middle ages up until sixty years ago.
29. Cow urine used to restart a biogas plant that has "stalled" when the bio-composting won't start or fizzles out in mid process then it can be jump started by adding ammonia (a BI -product of urine).
30. The dung ashes make excellent tooth power.
31. A good "cement" is made by mixing dung with water and mixing with dirt. The salt in the dung keeps the moisture in the "cement " from drying out as easily and the "Cement" feels cool and almost damp and lowers the temp in a well insulated house by ten or twelve degrees.
32. Moulding fresh pies into various shapes like rabbits and frogs and selling them as art / fertilization art. Place in a flower pot and a little fertilizer leaches out with each watering and the Poo pet slowly dissolves into the soil. and fresh cow dung assures that flies will not settle there.

33. If you make a biogas composer you can cook off the gas from the dung then when all the gas has been used up, you still have excellent compost in liquid form to place on your garden.
34. To break the parasite infestations whenever possible collect and place all dung and urine from all animal and human sources in a bio-gas generator. The digester turns the dung into sterile compost all worms et killed, and as a by product produces a gas called methane which is half as hot as propane but you can still use it to cook or run a generator to produce electricity. It is free and cleans up the area the finished product is essentially sterile and can be spread on crops as an excellent fertilizer.

The above list is by no means exhaustive and can still be added on to by imaginative and innovative cow-lovers and other persons. The primary uses of cow or cattle dung can be said to be for organic farming on one hand and on burning of cow-cakes for fuel on the other. There are also other important uses of cattle dung such as for production of bio-gas etc.

Conclusion:

Livestock contribute generously in today's world with diversified importance of their varied by- products. Most often, livestock is the only source of cash income for subsistence farms and also serves as insurance in the events of crop failure. As present demand for livestock products and by- products are increasing, it requires improved livestock and their productivity. We depend almost for everything on livestock products and by-products starting from food, clothing, and various accessories to medicinal needs. Hence, along with livestock products, by-products of livestock should be encouraged.

References:

Pandey, R.K. and Singh, D.(2004).Milk production in India: A cost-benefit Analysis. Ganga Kaveri Publishing House D.35/77, Jangamawadimath, Varanasi-221001.

Internet sources:

http://aitc.oregonstate.edu/resources/pdf/activity/cattle_by.pdf 10 FEB 2011

http://jas.fass.org/cgi/reprint/68/12/4200.pdf 10 FEB 2011

http://www.epa.gov/oecaagct/ag101/porkproducts.html 10 FEB 2011

http://dahd.nic.in/ch52/chap5.2.htm#11 11 FEB 2011

http://www.nddb.org/statistics/population_india_species.html 12 FEB 2011

http://www.mifugo.go.tz/documents_storage/Promoting%20other% 31 MAY 2011

* * *

21

ECONOMIC EVALUATION OF SETTLED AND SHIFTING CULTIVATION ALONG WITH THEIR FARMING SYSTEMS IN THE HILL ZONE OF ASSAM

D.C. KALITA AND B.C. BHOWMICK[1]
Department of Rural Development and Agricultural Production,
North-Eastern Hill University, Tura Campus, Meghalaya-794002, India

ABSTRACT

The present study was undertaken in the hill zone of Assam which comprised of Karbi Anglong and North Cacher Hill districts. Both the production systems of settled and shifting cultivation are followed in the zone. The study was based on primary data of 200 sample households out of which 100 sample households followed shifting cultivation and the rest 100 sample households followed the settled cultivation. The sample households were selected by using multistage random sampling procedure and selected households were categorized into three size groups based on area under settled and shifting cultivation using cumulative root frequency rule. Data pertained to the year 1999-2000 and were collected from the sample households by interview method using structured schedules. The farming systems under settled and shifting cultivation were identified based on crop and livestock activities. The economic evaluation of the cultivation system as well as the farming systems were done mainly by estimating the output-input ratio. Results showed that there were 17 farming systems under settled cultivation as against 23 farming systems under shifting cultivation. The economic evaluation showed that the output-input ratio was higher in shifting cultivation than the settled cultivation. However, the net return was found to be higher in settled cultivation than that of shifting cultivation. Besides the output-input ratio in all the farming systems identified under settled and shifting cultivation was greater than equity.

Introduction:

The agricultural production systems in the hilly areas differ from the plough cultivation in the plain areas. In one side highly modern agricultural cultivation practices flourishes while on other side age old agricultural practices dominate. Thus dualism in agriculture still persists in the hilly areas. This system of dualism determines the standard of living of people. Further, there are two distinct agricultural production systems in the hilly areas, viz., settled and shifting cultivation. The tribal people practice the shifting cultivation which is locally known as *jhuming* and this system is the first step of transition from hunting to food production system. Usually the settled cultivation is practiced in foothills and terraces' in gentle slopes. The production behaviour in both the system is more or less similar as crop mixtures are grown in both the production system. However, the productivity under settled cultivation is higher than that of shifting cultivation (Chauhan, 2001). This productivity differences lead to differences in livelihood standard of people. Therefore, it is necessary as well as important to know the economics under both the production systems.

Farming system is recognized as the organization of agricultural activities into a unit for profitability harnessing the solar energy (Thakur, 1995). Further the farming systems are location specific. Therefore, one farming system cannot be recommended for other geographical area without modification to suit the local need condition. The study of the system as well as the farming systems will provide a broad basis for developing farm plans which in turn help the farmer to utilize his resources optimally for achieving sustainability to ensure a minimum standard of living. In the light of increase population in the state of Assam, promotion of sustainable agriculture is the only solution to feed the people. Sustainable agriculture is that form of farming which produces sufficient food to meet the needs of the present generation without eroding the ecological assets and the productivity of the life supporting system of future generation (Thakur, 1995). Keeping view the above facts the present study is an attempt to study the economics of settled and shifting cultivation along with their farming systems in hill zone of Assam.

Methods and Materials:

The present study was undertaken in the hill zone of Assam which comprised of Karbi Anglong and North Cacher Hill districts. Both the production systems of settled and shifting cultivation are followed in the zone. The study was based on primary data of 200 sample households out of which 100 sample households followed shifting cultivation and the rest 100 sample households followed settled cultivation. The sample households were selected by using multistage random sampling procedure and selected farmers were categorized into three size groups based on area under settled and shifting cultivation using cumulative root frequency rule (Cochran, 1977). Data pertained to the year 1999-2000 and were collected from the sample households by interview method using structured schedules. The economic evaluation of settled and shifting cultivation was done mainly by estimating the output-input ration along with the net return. The existing farming systems under settled and shifting cultivation were identified based crop and livestock activities (Bhowmick, *et. al.*, 1992). Under settled cultivation the combination of cropping sequences as practiced by the farmers together with various livestock activities reared in the homestead was considered as the prevailing farming system. On the other hand, different crop mixtures used by the sample farmers in shifting cultivation in the hill slopes together with various livestock activities reared in the homestead was considered as the prevailing farming system under shifting cultivation. Estimating the output-input ratio did the evaluation of different identified farming systems under settled and shifting cultivation.

Results and Discussion:

Economic evaluation of settled and shifting cultivation:

The cost and return of different cropping sequences under settled cultivation and different crop mixtures under shifting cultivation was computed separately and then total production cost and total gross return were estimated and finally net return and output-input ratio was estimated. The results of the cost and return under settled and shifting cultivation are presented in Table 1 and 2. The Tables revealed that output-input ratio in both settled and shifting cultivation were more than equity. However, the estimated output-input ratio in shifting cultivation was found to be higher (2.18) as compared to settled cultivation (1.77). Saikia and Bora (1971) also reported that the return per unit of land from mixed crops was much higher then in shifting cultivation than that terrace cultivation in hilly areas of north east India. But considering the ecological effect of shifting cultivation this higher ratio should not be permanent and considered to be better than the settled cultivation. The variation in output-input ratio amongst the size groups of farm was due to difference in variable costs incurred by the sample farmers. However, the net return per hectare under settled cultivation was higher than that of shifting cultivation. This might be due higher productivity of crops especially paddy in settled cultivation compared to shifting cultivation. Chauhan (2001) reported that yield of paddy in shifting cultivation was 15.20 per cent less as compared to settled cultivation due to agroeconomic consequences of the system.

Identification of cropping sequence and crop mixtures under settled and shifting cultivation:

Among the various cropping sequence raised by the sample farmers under settled cultivation, CS 10 (*ahu* rice-*sali* rice-fallow) was the major cropping sequence followed by about 18% of the total sample farmers and accounting 37.46% of total cropped area. However CS 3 was the major cropping sequence followed in upland in terms of area coverage. In medium land CS-7 was the major cropping sequence followed by 9% of the sample farmers (Table3).

Crop mixture IV (rice + ginger + vegetables + turmeric + chilli + sesamum + mustard) was the dominant crop mixture under shifting cultivation adopted by 29% of the total farmers. This was followed by rice + maize + ginger+ vegetables + colocasia + marua (25%) and rice + ginger + vegetables + pumpkin + cotton + okra (23 %), and rice maize + ginger + vegetables + colocasia + pumpkin (23%) respectively (Table 4). The major difference occurred mainly on inclusion of one or more crops in the mixtures.

Identification of farming systems under settled and shifting cultivation:

The identified major farming systems under settled and shifting cultivation are presented in Table 5 and 6. The Table showed that FSST 17 was the dominant farming system which was followed by 18 % of the sample farmers. FSST 17 farming system was followed by FSST 14 (11%) and FSST 11 (8%). Amongst the different size groups of farm, FSST 17 was also the dominant farming in all the size groups accounted 18.42, 12.50 and 23.33 pr cent in group I, II and III farms respectively.

Among the identified existing farming systems under shifting cultivation, FSSH 17 was the dominant farming system followed by about 13% of the total sample farmers. The farming system was followed by FSSH 12 and FSSH 21 (8%). FSSH 22 was also the dominant farming system (12.50%) in group-I farms, FSSH 17 in group-II (17.50%) and FSSH 12 in group-III farms.

Evaluation of farming systems under settled and shifting cultivation:

The evaluation farming systems in vague in the study area was mainly done by using output-input ratio and are presented in Table 7 and 8. The Table revealed that all the farming systems in vague was barely viable (economically) since the estimated output-input ratio could be lower than equity upon inclusion of all costs in estimating output-input ratio. FSST 2 farming system (CS 2 + poultry + goattery) generated the highest (1.46) output-input ratio amongst the farming systems under settled cultivation.

Kalita (1995) reported the similar findings in case of crop + poultry + goattery farming system in Sonitpur district of Assam.

Similarly all the identified farming systems under shifting cultivation were also barely viable like settled cultivation. Similar findings were reported by Kalla (1988). Amongst the farming systems under shifting cultivation, FSSF 16 and FSSF 23 generated the highest (1.69) output-input ratio and FSSF 13 generated the lowest (1.25) output-input ratio. The highest output-input ratio in farming systems was mainly due to higher relative share of income from crop mixture IV.

Conclusion:

The following conclusions could be drawn from the above results and discussions.

1. The output-input ratio was higher in shifting cultivation than that of settled cultivation. However the per hectare net return was higher in settled cultivation than that of shifting cultivation.
2. All the farming systems identified under settled and shifting cultivation was barely viable.

Policy Implication:

The present study revealed that the deep rooted shifting cultivation is the main agricultural production system in the hilly areas. An attempt should be made to update the traditional production system with modern technology with appropriate technological intervention. The crop mixture IV (rice + ginger + vegetables + turmeric + chilli + sesamum + mustard) could be suggested under shifting cultivation for better economic return.

References:

Bhowmick, B.C.; Mazumder, D.K. and Das, R., 1992. Identification and Optimization of Resources in some Major Farming system. *Agricultural Economic Research Review*. 5(1):72-90

Chauhan, B.S., 2001. Shifting Cultivation in Perspective. Nagaland University Publication. Kohima, Nagaland. Pp-4

Cochran, G.W., 1977. Sampling Technique. 3rd Edition. Wiley Eastern Limited. New Delhi.

Kalita, P., 1995. Impact of Diversification and Liberal Credit Policy on Income and Employment of Non-viable Farmers in Sonitpur District of Assam. Unpublished M.Sc. Thesis Presented to Assam Agricultural University.

Kalla, J.C., 1988. Forestry and Economics Development. *Indian journal of Agricultural Economics.* 43(3):237-255

Saikia, P.D. and Bora, D., 1971. Pattern of Crop Production under Shifting Cultivation and Settled Cultivation – A study in Garo Hills, Meghalaya. *Indian journal of Agricultural Economics.* 43

Thakur, A. C., 1995. Farming system research for Sustainable Agriculture. Proceedings of Seminar of Agricultural science Society, North East India. 86-99

Table 1: Cost and Return of Settled Cultivation in various Size Groups of Farm

Sl.No.	Particulars	Group-I	Group-II	Group-III	All
1.	1. Total production cost	5987	6101	6403	6168
2.	2. Total gross return	10721	10916	11140	10925
3.	3. Net return	4734	4815	4737	4757
4.	4. Output-input ratio	1.79	1.79	1.74	1.77

Table 2: Cost and Return of Shifting Cultivation in various Size Groups of Farm

Sl.No.	Particulars	Group-I	Group-II	Group-III	All
1.	Total production cost	8026	9444	8579	8683
2.	Total gross return	12104	14819	13241	13387
3.	Net return	4078	5375	4662	4704
4.	Output-input ratio	2.03	2.32	2.19	2.18

Table 3: Identification of Cropping sequences Across Various Size Groups under Settled Cultivation

Cropping sequence	Group-I		Group-II		Group-III		All	
	Number	Area	Number	Area	Number	Area	Number	Area
A. Upland:								
1. Ginger	4 (10.53)	0.35 (3.31)	3 (9.38)	1.18 (3.67)	5 (16.67)	4.95 (9.18)	12 (12.00)	2.16 (6.68)
2. Turmeric	3 (7.89)	0.42 (3.97)	4 (12.50)	1.23 (3.78)	3 (10.00)	2.43 (4.50)	10 (10.00)	1.36 (4.21)
3. Ahu rice-sesamum-mustard	5 (13.16)	1.81 (17.14)	2 (6.25)	2.75 (8.46)	2 (6.67)	3.33 (6.17)	9 (9.00)	2.63 (8.17)
4. Ahu rice-fallow-mustard	3 (7.89)	0.29 (2.75)	3 (9.38)	2.15 (6.62)	2 (6.67)	2.30 (4.26)	8 (8.00)	1.58 (4.87)
5. Fallow-sesamum-mustard	2 (5.26)	0.67 (6.34)	3 (9.38)	1.40 (4.31)	3 (10.00)	1.36 (2.52)	8 (8.00)	1.14 (3.57)
B. Medium land:								
6. Ahu rice-maize-fallow	2 (5.26)	0.78 (7.39)	3 (9.38)	3.41 (10.49)	2 (6.67)	5.00 (9.27)	7 (7.00)	3.06 (9.46)
7. Ahu rice-fallow-vegetables	4 (10.53)	1.20 (11.36)	3 (9.38)	3.04 (9.35)	2 (6.67)	3.33 (6.17)	9 (9.00)	2.52 (7.79)
8. Colocasia	3 (7.89)	0.50 (4.73)	3 (9.38)	1.44 (4.43)	2 (6.67)	2.54 (4.71)	8 (8.00)	1.49 (4.61)
C. Low land:								
9. Fallow – Sali rice – fallow	5 (13.16)	1.57 (14.87)	4 (12.50)	5.60 (17.23)	2 (6.67)	5.65 (10.47)	11 (11.00)	4.27 (13.21)
10. Ahu rice – Sali rice - fallow	7 (18.42)	2.97 (28.13)	4 (12.50)	10.30 (31.69)	7 (23.33)	23.06 (42.74)	18 (18.00)	12.11 (37.46)
Total	38 (100)	10.56 (100)	32 (100)	53.95 (100)	30 (100)	53.95 (100)	100 (100)	32.33 (100)

Figures in parentheses indicate percentages to total sample farmers and gross cropped area

Table 4: Identification of Crop Mixtures Across Various Size Groups under Shifting Cultivation

Crop Mixtures	Name of Crop Mixture	Group I	Group II	Group III	All
Rice + maize + ginger + vegetables + colocasia + marua	CM I	8 (25.00)	11 (27.50)	6 (21.43)	25 (25.00)
Rice + maize + ginger + vegetables + colocasia + pumpkin	CM II	9 (28.13)	9 (22.50)	5 (17.86)	23 (23.00)
Rice + maize + ginger + vegetables + colocasia + pumpkin + cotton + okra	CM III	6 (18.75)	10 (25.00)	7 (25.00)	23 (23.00)
Rice + ginger + vegetables + turmeric + chilli + sesamum + mustard	CM IV	9 (28.13)	10 (25.00)	10 (35.71)	29 (29.00)
Total		32 (100)	40 (100)	28 (100)	100 (100)

Figures in parentheses indicate percentages to total sample farmers

Table 5: Identification of Farming Systems Across Various Size Groups under Settled Cultivation

Farming system	Name of the farming system	Group I	Group II	Group III	All
1.CSI+dairy+buffalo+poultry+goattery+ piggery	FSST 1	2 (5.26)	2 (6.25)	1 (3.33)	5 (5.00)
2. CSI +poultry+piggery	FSST 2	-	-	1 (3.33)	1 (1.00)
3. CSII+dairy+goattery+piggery	FSST 3	1 (2.63)	1 (3.13)	1 (3.33)	3 (3.00)

4. CSII+poultry+goattery	FSST 4	2 (5.26)	2 (6.25)	3 (9.99)	7 (7.00)
5. CSIII+dairy+poultry+duckery+piggery	FSST 5	-	4 (12.50)	2 (6.67)	6 (6.00)
6. CSIII+buffalo+goattery+piggery	FSST 6	5 (13.16)	-	1 (3.33)	6 (6.00)
7. CSIV+poultry+goattery+piggery	FSST 7	3 (7.89)	2 (6.25)	2 (6.67)	7 (7.00)
8. CSIV+ piggery	FSST 8	-	2 (6.25)	-	2 (2.00)
9. CSV+buffalo+piggery	FSST 9	2 (5.26)	2 (6.25)	2 (6.67)	5 (5.00)
10. CSV+poultry+goattery	FSST 10	1 (2.63)	-	2 (6.67)	3(3.00)
11. CSVI+dairy+poultry+piggery	FSST 11	4 (10.53)	2 (6.25)	-	8 (8.00)
12. CSVII+poultry+piggery	FSST 12	-	2 (6.25)	2 (6.67)	2 (2.00)
13. CSVIII+poultry+goattery+piggery	FSST 13	3 (7.89)	1 (3.13)	3 (9.99)	6 (6.00)
14. CSIX+dairy+buffalo+goattery+ piggery	FSST 14	5 (13.16)	3 (9.38)		11 (11.00)
15. CSX+dairy+poultry+goattery	FSST 15	3 (7.89)	1 (3.13)	-	2 (2.00)
16. CSX+poultry+duckery+piggery	FSST 16	-	2 (6.25)	2 (6.67)	4 (4.00)
17. CSX+poultry+piggery	FSST 17	1 (18.42)	4 (12.50)	7 (23.33)	18 (18.00)
Total	17	38 (100)	32 (100)	30 (100)	100 (100)

Figures in parentheses indicate percentages to total sample farmers

Table 6: Identification of Farming Systems Across Various Size Groups under Shifting Cultivation

Farming system	Name of the farming system	Group I	Group II	Group III	All
1. CMI + dairy	FSSH 1	2 (6.25)	1 (2.50)	-	2 (2.00)
2. CMI + poultry + buffalo	FSSH 2	-	1 (2.50)	2 (2.74)	3 (3.00)
3. CMI I+ dairy+ buffalo + goattery	FSSH 3	-	3 (7.50))	-	1 (1.00)
4. CMI +buffalo	FSSH 4	-	4 (10.00)	2 (2.74)	5 (5.00)
5. CMI+ buffalo + poultry+ goattery	FSSH 5	1 (3.13)	-	1 (3.57)	6 (6.00)
6. CMI+buffalo+goattery+poultry+ piggery	FSSH 6	3 (9.38))	1 (2.50)	1 (3.57)	4 (4.00)
7. CMI+poultry+ piggery	FSSH 7	3 (9.38))	2 (5.00)	-	4 (4.00)
8. CMII + dairy + buffalo + goattery + poultry+piggery	FSSH 8	1 (3.13)	1 (2.50)	-	2 (2.00)
9. CMII+buffalo+goattery	FSSH 9	1 (3.13)	1 (2.50)	-	3 (3.00)
10. CMII+poultry+goattery+piggery	FSSH10	-	2 (5.00)	-	31 (1.00)
11.CMII+buffalo+goattery+poultry+ piggery	FSSH11	2 (6.25)	1 (2.50)	1 (3.57)	3 (3.00)
12. CMII+poultry	FSSH12	2 (6.25)	1 (2.50)	4 (14.29)	8 (8.00)

13.CMII+poultry+ piggery	FSSH13	2 (6.25)	4 (10.00)	-	4 (4.00)
14.CMIII+dairy+buffalo+poultry+goattery+piggery	FSSH14	2 (6.25)	2 (5.00)	-	3 (3.00)
15. CMIII+goattery+poultry+piggery	FSSH15	2 (6.25)	2 (5.00)	-	3 (3.00)
16. CMIII+buffalo	FSSH16	-	1 (2.50)	3 (10.71)	13 (13.00)
17. CMIII +poultry +piggery	FSSH17	3 (9.38))	-	3 (10.71)	2 (2.00)
18. CMIV +buffalo	FSSH18	-	7 (17.50)	-	4 (4.00)
19. CMIV+buffalo+goattery	FSSH19	-	2 (5.00)	2 (2.74)	6 (6.00)
20. CMIV+ goattery+poultry+ piggery	FSSH20	2 (6.25)	2 (5.00)	2 (2.74)	8 (8.00)
21. CMIV+poultry+piggery	FSSH21	3 (9.38))	3 (7.50)	2 (2.74)	7 (7.00)
22. CMIV+ piggery	FSSH22	4 (12.50)	-	3 (10.71)	2 (2.00
23. CMIV+ goattery+poultry	FSSH23	-	-	2 (2.74)	2 (2.00)
Total	23	32 (100)	40 (100)	28 (100)	100 (100)

Figures in parentheses indicate percentages to total sample farmers

Table 7: Results of the Output-Input Ratio of various Identified Farming Systems Across Size Groups of Farm under Settled Cultivation

Farming system	Group-I	Group-II	Group-III	All
FSST 1	1.38	1.43	1.42	1.41
FSST 2	-	-	1.41	1.41
FSST 3	1.31	1.45	1.41	1.39
FSST 4	1.47	1.51	1.39	1.46
FSST 5	-	1.40	1.40	1.40
FSST 6	1.24	-	1.42	1.33
FSST 7	1.37	1.44	1.43	1.41
FSST 8	-	1.45	-	1.45
FSST 9	1.35	1.38	1.39	1.37
FSST 10	1.43	-	1.34	1.39
FSST 11	1.32	1.38	1.36	1.35
FSST 12	-	1.38	-	1.38
FSST 13	1.36	1.43	1.42	1.40
FSST 14	1.43	1.43	1.46	1.44
FSST 15	1.45	1.44	-	1.45
FSST 16	-	1.37	1.40	1.39
FSST 17	1.34	1.38	1.40	1.37

Table 8: Results of the Output-Input Ratio of various Identified Farming Systems Across Size Groups of Farm under Settled Cultivation

Farming system	Group-I	Group-II	Group-III	All
FSSH 1	1.66	-	-	1.66
FSSH 2	-	1.35	1.49	1.42
FSSH 3	-	1.35	-	1.35
FSSH 4	-	1.55	1.59	1.57
FSSH 5	1.58	1.52	1.58	1.56
FSSH 6	1.41	-	1.51	1.41
FSSH 7	1.43	-	-	1.43
FSSH 8	1.30	1.27	-	1.29
FSSH 9	1.31	1.35	-	1.33
FSSH10	-	1.26	-	1.26
FSSH11	1.31	1.38	1.42	1.37
FSSH12	1.42	1.58	1.43	1.48
FSSH13	1.27	1.23	-	1.25
FSSH14	1.42	1.27	-	1.35
FSSH15	1.50	1.24	-	1.37
FSSH16	-	-	1.69	1.69
FSSH17	1.47	1.44	1.49	1.47
FSSH18	-	1.42	-	1.42
FSSH19	-	1.40	1.47	1.44
FSSH20	1.22	1.25	1.44	1.30
FSSH21	1.66	1.24	1.42	1.44
FSSH22	1.84	-	1.41	1.63
FSSH23	-	-	1.69	1.69

22

MEAT CONSUMPTION PATTERN AMONG MAJOR TRIBES IN NAGALAND, INDIA

N. K. PATRA[1], M. N. ODYUO[1], N. GHOSH[2], CH. VICTORIA DEVI[3] AND S. MONDAL[3]

[1] Department of Agricultural Extension, School of Agricultural Sciences and Rural Development, Nagaland University, Nagaland – 797106, India.

[2] Department of Animal Science and

[3] Department of Agricultural Extension, Bidhan Chandra Krishi Viswavidyalaya, Mohanpur, Nadia, West Bengal – 741252, India Corresponding author's E-mail:gnilotpal@yahoo.com

ABSTRACT

People of Nagaland are very much fond of meat and they eat various types of meat. The present study was conducted in two villages under two districts of Nagaland, India. In this study four major tribes namely, Konyak, Ao, Lotha and Angami were considered and 100 respondents from Konyak, 55 from Ao, 40 from Lotha and 40 from Angami tribes were interviewed. Five popular sources of meats viz., pork, mithun meat, dog meat, chicken and beef were identified and presented in 10 possible pairs (combinations) to the respondents and they were asked to select one type of meat over the other from each pair separately, which they considered better from quality point of view.as well as from taste point of view. For the both sets of study the 'Method of Paired Comparisons' were followed separately to find out the relative preferences of meat relating to its quality and taste parameter only. The preference of meat from quality point of view as per scale value was as follows: pork is the highly preferred meat and dog meat is the least preferred meat. Chicken occupied the 2nd position in case of all the tribes studied. The preference of meat according to taste as per scale value was more or less same and study revealed that dog meat losing its popularity.

Introduction:

Rearing of cattle (*Bos indicus*), chicken (*Gallus domesticus*) and pig (*Sus domesticus*) are the common and conventional enterprises in rural as well as urban areas throughout the State of Nagaland, India. Primary purpose of rearing these animals is mainly for domestic consumption and also for marketing of surplus meat. Another four footed domestic animal is dog (*Canis familiaris*) and purpose of domestication is mainly for pleasure, companionship and to some extent for security also. But in North Eastern Region of India especially in Nagaland, apart from the above mentioned reasons for domestication of dog, another important aspect is its quality meat. Mithun (*Bos frontalis*) is also an important ruminant animal, semi domesticated to wild in nature and it is an important source of meat in Nagaland.

Traditionally people are rearing the pigs and fowls (poultry) as homestead activities at the surroundings of their household. Shed structure for piggery is present, and for poultry birds it is not present in some cases. In rural areas poultry birds are staying on tree at night time or in pig sty. Dog rearing is generally continued inside the household and in few cases separate kennel is present for this purpose. Cattle and mithum rearing are totally based on grazing. Cattle are generally kept in the sheds outside their residential areas; whereas mithun is preferred to stay in the nearby forest.

Domesticated animals in rural areas are mainly of *deshi*/local variety or breed. Crossbreds are also available in case of dog and poultry birds; cattle and pigs are purely of indigenous type. Mithuns are indigenous to Nagaland, however, crosses of Nagaland and Arunachal type mithuns are also available. This crossing is natural without human interference.

Food habit of traditional people of the State is unique and they generally prefer boil curry of their choices. Naga people (traditional inhabitants of Nagaland) are fond of meat but they do not prefer spicy and oily preparations. The people have their own products to improve the taste and quality of meat and vegetables curry. One of the prominent indigenous technical knowledge (ITK) of local people is bamboo shoot fermented liquid and they used to add it in all types of curry irrespective of vegetable and non vegetable based preparations.

Objectives of the Study:

The main objective of this study is to find out the hierarchy of meat preference among different tribes of Nagaland with respect to quality and taste of meat of pork, chicken, beef, mithun,and dog which are consumed by the major tribes in Nagaland and also the reasons of meat preference among the tribes. In this study, quality and taste of meat have been judged by the respondent's choice and without consideration of scientific basis. Traditional people have different believes and norms in every respect which directly related to preference and consumptions of food. Taste is palatability and degree of pleasure owing to consumption of any consumable food and liquid substances which satisfy our sensory parts inside mouth and it differs from person to person. The hierarchical preferences from quality and taste of different types of meats shall provide a clear picture about the requirements of the meats in a particular area, which may be helpful for planners, policy makers, researchers and businessmen for planning, policy formulation and maintaining proper supply as per demand.

Methodology:

Nagaland State was inaugurated as the sixteenth State of India on 1st December,1963. It is bounded by Assam in the West, Myanmar on the East, Arunachal Pradesh and part of Assam on the north and Manipur in the South (Raatan, 2004). The State consists of eleven Administrative District, Inhabited by sixteen major tribes *viz.*, Konyak, Aou/Ao, Lotha, Angami, Sema, Rengma, Chakesang, Sangtam, Phom, Chang, Yimnhunger,

Khiamniungam, Zeliang, Kuki, Pochury *etc*, along with many sub-tribes. Each tribe has its own custom, language and dress. Nature could not have been kinder to Nagaland, sometimes referred to as the 'Switzerland of the East', the exquisitely picturesque landscapes, the vibrantly colourful sunrise and sunset, lush and verdant flora-this is a land that presents unimaginable beauty, moulded perfectly for a breathtaking experience.

The state approximately lies between $25^0 6'$ and $27^0 4'$ latitude; North of Equator and between the Longitudinal lines $93^0 20'$and $95^0 15'$E. Geographical area is 16527 sq km and total population is 1,988,636 (as per 2001 census). Density of population is around 120 per sq km. Average annual rainfall ranges from 2000-3000 mm and temperature ranges from 4^0 to 31^0C.

Selection of Respondents:

Two districts of Nagaland, namely Dimapur and Mon were purposively selected for the present investigation. A total of 235 numbers of tribal people were considered as respondents irrespective of sex for the present study and they belong to Konyak (N=100), Aou/Ao (N=55), Lotha (N=40) and Angami (N=40) tribes. The Dimapur district is inhabited by all major tribes *viz.*, Aou/Ao, Lotha, Angami *etc.*, whereas Mon District is purely for Konyak tribes. From Mon District only one village was selected for this study and 100 villagers were randomly selected as respondents. In Dimapur district two villages were selected and a total of 135 respondents were randomly considered for the study. All the respondents were subdivided according to their tribes. All the respondents were personally interviewed through an Interview Schedule. The data were collected from August to December, 2009.

The Method of Paired Comparisons:

Paired comparison is a psychological scaling method and makes possible the quantitative investigation of all kinds of values and subjective experience (Edwards, 1969). In this method, the stimuli (items, statements or variables) are presented in pairs in all possible combinations and the respondents are asked to select one stimulus over the other from each pair, which is judged as more favourable. Because the stimuli are presented in pairs and one stimulus is compared over the other, this is known as *the Method of Paired Comparisons*. This method of psychological scaling also provides an estimate of the distances between each of the stimuli, in comparison to the stimulus with least preference, whose scale value is (arbitrarily) brought down to the level of 'zero'. If there are 'n' stimuli, the numbers of pairs which may be obtained are n (n – 1)/2. For example, if there are 4 stimuli, the number of pairs will be 4(4 – 1)/2=6. The stimuli or items for judgement should be distinct from each other and easily understandable. To eliminate response bias, both the stimuli in each pair and the pairs themselves, are randomly arranged. The stimuli are then presented to the respondents, who are asked to select one stimulus over the other from each pair, which they consider as more favourable. The stimuli selected are appropriately marked.

The F-matrix:

The first table shall consist of *frequencies* correspond to the number of times that each stimulus is judged more favourable than the other. The cell entries corresponding to the frequency with which the column stimulus is judged more favourable than the row stimulus.

The P-matrix:

The P-matrix gives the *proportion* of times the column stimulus is judged more favourable than the row stimulus. This is obtained by dividing each of the cell entries in the F-matrix by N, *i.e.*, total number of respondents. The cell entries in the diagonal line which are blank in the F-matrix are assumed to be N/2. This again divided by N, to get the proportion of 0.500 in each cell of diagonal line in the P-matrix.

The Rearranged P-matrix:

The rearranged P-matrix is then made with the stimulus having the smallest column sum at the left and that with the highest at the right.

The Z-matrix:

The Z-matrix gives the *normal deviates* corresponding to the proportions in the table of P-matrix. The Z-matrix corresponding to the rearranged P-matrix is obtained by converting the pij entries to zij entries with the help of the table of Normal Deviates (Edwards, 1969). The column sum (Sums z) for each stimulus is obtained by adding the respective cell entries, taking the sign into consideration. The means (Mean z) are obtained by dividing the sums with the number of stimuli in each column. The absolute scale value of the stimulus with the largest negative deviation is added to all the column means to make the scale value for this stimulus zero and all others with positive sign.

Five commonly used meat, *viz.* (A) pig meat (pork), (B) mithun meat, (C) dog meat, (D) fowl meat (chicken) and (E) cattle meat (beef) were identified in consultation with local people. These were arranged in 10 possible pairs (Table 1) and presented to the respondents to select one meat over the other from each pair separately, which they considered to be more preferable from quality point of view. The respondents were also asked to judge the meat from each pair separately which they considered more tasty. In this study, quality and taste are considered according to the respondents choice and without consideration of scientific basis. Quality of meat was indicative of juiciness, tenderness, easy to digest, any concern to quick recovery of health problem, and any other social and traditional belief in their system. Taste is palatability and degree of pleasure due to consumption of any consumable food and liquid substances which satisfy our sensory parts inside mouth and it differs from person to person. After the separate personal interview for both the sets of study, all the respondents were grouped according to their tribes in both the sets of studies. Later the Method of Paired Comparisons (Edwards, 1969) has been adopted separately in both the sets of studies *i.e.*, two different sets of analysis for respective tribes, namely, for Konyak, Aou/Ao, Lotha and Angami.

In paired comparisons, the method of complete data was used as there was no Pij value equal to or greater than 0.99 or equal to or less than 0.01. For analysis of data, F-matrix which consist of frequencies corresponding to the number of items that each column stimulus was judged more favourable than the row stimulus, followed by P-matrix, rearranged P-matrix and Z-matrix of five different sets of analysis were computed.

Table 1. Ten pairs of meat used for the study

1	(A) Pork (B) Mithun meat	6	(A) Pork (D) Chicken
2	(C) Dog meat (E) Beef	7	(B) Mithun meat (C) Dog meat
3	(A) Pork (C) Dog meat	8	(C) Dog meat (D) Chicken
4	(D) Chicken (E) Beef	9	(B) Mithun meat (E) Beef
5	(B) Mithun meat (D) Chicken	10	(A) Pork (E) Beef

Results and Discussion:

Analysis of two different sets of data relating to 'meat preference as per quality of the meat' and 'meat preference as per taste of the meat' among four different tribes has been considered. The Z-matrix of meat preferences of each tribe has been presented in Table 2.a, 3.a, 4.a and 5.a for determination of quality and Table 2.b, 3.b, 4.b and 5.b for determination of taste of meat respectively.

Pork

The present study reveals that from quality point of view, pork has the highest scale value in case of all tribes; namely, Konyak (Table 2.a), Ao (Table 3.a), Lotha (Table 4.a), and Angami (table 5.a). Study also shows that, pork has again been occupied highest position among all tribes (table 2.b, 3.b, 4.b and 5.b)according to taste Pork has been considered as the first choice and most frequently consumed meat among all the tribes in Nagaland. Pork is the culinary name of meat from the domestic pig. In Nagaland people generally purchase meat from the market and consume fresh meat after cooking and at the same time they put excess amount of meat onto certain height of their traditional oven for smoke treatment, mainly for preservation and improvement of taste and quality. Pork is the most commonly and widely consumed meat throughout the world. Pork is contributing around 38 per cent of world meat production and trend of consumption is upward in direction. According to the USDA's Foreign Agricultural service, nearly 100 million metric tons of pork were consumed worldwide in 2006. However, consumption and pattern of consumption vary widely from place to place and nation to nation. In India, consumption of pork in Hinduism is not so common and in Islamic religion it is totally prohibited. In North-Eastern States of India especially in Nagaland, more than 90% of population is Christian by religion, and social and religious prohibition is not present to consume any types of meat. Here trend of consumption and preference is as usual and similar to world trend (Govt. of Nagaland, 2006).

Chicken

Preference of chicken is second in position among all the tribes in case of both sets of analysis. But differences in scale value compared to first preference exist in case of quality and test point of view. In case of Konyak tribes (Table 2.a) taking into consideration of **quality** of meat the difference (scale value) between 1st and 2nd preferences is only **0.088** (1.610-1.522); where in case of **taste** (table 2.b) the difference is **0.185** (1.437-1.252). In Ao tribes (Table 3.a) difference is **0.315** (0.9 94-0.679) in **quality** study where **taste** (table 3.b) difference is **0.764** (1.599-0.835). Same way, in case of Lotha tribes (Table 4.a and table 4.b) and Angami tribes (Table 5.a and table 5.b) scale value difference between 1st and 2nd preference in respect of **quality** is **0.277** (1.250-0.973) and **0.287** (1.086-0.799) respectively, and in respect of **taste** in Lotha tribes difference is **0.359**(1.844-1.485) and in case of Angami tribes differences is **0.323** (1.283-0.960) these are considered as highly similar preference group.

In this respect preference of chicken against pork of Konyak tribes from quality and taste point of view differences are negligible (0.088 and 0.185 respectively) and it may be concluded that chicken is preferred in terms of quality and taste point of view by Konyak tribes.

In case of Ao tribes' differences in preference of chicken against pork from quality and taste point of view are wide (0.315 and 0.764) and preferential difference (quality and taste) is also higher compare to Konyak tribes. As a result, in case of Ao tribes, degree of preference chicken compare to pork is relatively less and in comparison to Konyak tribes also less.

In case of Lotha tribes, the difference of preference of chicken from quality point of view against pork is 0.277, which also lower than the difference in case of Ao (0.315) and near to similar with Amgami tribes' preferential difference (0.287). From taste point of view scale difference between 1st and 2nd preference is 0.359 in case of Lotha tribes and 0.323 in case of Angami tribes. So preference of chicken relating to taste in case of Lotha and Angami tribes is very nearer. Study also revealed that these two tribes preferred chicken with respect to quality and taste.

Negligible differences in scale value between pork and chicken preferences in Konyak tribes may be attributed to their economic conditions. Study area for the Konyak tribes is extremely interior village of Mon district and their poverty is prominent. Accessibility of essential and daily requirement is also difficult. Side by side whole pig is a costly and finished product is enormous in amount which is difficult to consume by a single family. Whereas, chicken is easily arranged for consumption in any time due to its low price per unit, and in maximum occasions it is available in their own domestic farms.

On the other hand, respondents of other tribal groups reside relatively progressive village areas of Dimapur district. In this area weekly market and daily market facility are present and different types of meats are available for purchasing. People are able to purchase meat as per their choice. It has been mentioned earlier that pork has got the first preference among all sources of meats studied and when it is available, then consumers also purchase as per their preference.

Mithun meat

In Table 2,a, mithun meat has got 3rd preference by Konyak tribes from quality point of view. Same way, it is 3rd in position in case of Ao tribes (Table 3.a). The scale difference in between 2nd and 3rd preference of Konyak tribes is around 0.268 (1.522-1.254); in case of Ao tribes, it is around 0.330 (0.679-0.349). In these two analyses *i.e.* 'Konyak', and 'Ao' scale value differences are 0.268 and 0.330 respectively and in increasing order. In case of Konyak scale value difference is lowest (0.268) compared to others. On the other hand mithun is the 5th preferred meat in Lotha tribes (Table 4.a) and 4th preferred meat in Angami tribes (Table 5.a) with quality point of view. Here study revealed that, taking into consideration of quality of meat, mithun meat is the least preferred meat among the all five types of meats in case of Lotha tribes from quality and taste point of view and in case of Angami tribe's it is the second least preferred meat from both the way.

Taking into consideration of taste of meat, Table 2.b shows that mithun meat is the 3rd preferred meat in case of Konyak tribes, and scale value difference between 2nd and 3rd preferred meat is 0.265. That is near to equal with scale value difference between 2nd and 3rd preferred meat from quality point of view. So, preference of mithun meat is very similar from quality and taste point of view by Konyak people. In case of Ao tribes' mithun meat preference from taste point of view is 4th in position and where quality point of view it is in 3rd preferred meat.

In this study area *i.e.*, Konyak village of Mon district, around 688 semi-domesticated mithun is present and villagers are the owner of all the mithuns (SASRD NAIP, 2008). Local availability of the source might be the reason for high preference and low difference of scale value.

Beef

Beef is the 4th preferred meat from quality point of view in case of all tribes, except Angami tribes (Table 5.a) where it occupied 3rd position among all the meats considered for this study. In case of Angami tribes in respect of quality of meat scale difference between 2nd and 3rd preference is 0.466 (0.799-0.333) which indicate that the people preferred but not so strongly. From taste point of view it is 3rd in position in case of Ao, Lotha and Angami tribes and 4th in position in case of Konyak tribes. In case of Lotha tribes scale difference from taste point of view, between 2nd and 3rd preference is1.166 (1.485-0.319) and extremely wide which also indicates the weak preference compared to 2nd preference.

During interview respondents has expressed their opinion as both the meats namely mithun meat and beef are fibrous in nature, and it is the main reason for low preference. In addition to this, after slaughtering of cattle and mithun, finished product is huge in amount and it is not suitable for sale in the small village market, where consumers are limited in number for purchasing the total product. As a result local slaughterers are avoiding slaughter these two meat animals.

Dog meat

In India, dog is a popular source of meat only in North-Eastern parts and mainly in Nagaland, Mizoram and to some extent in other States of the region. Most of the people from all tribes of Nagaland like to consume dog meat either occasionally or regular basis but preference of dog meat is gradually in decreasing order and very much prominent in case of Konyak tribes. In this study, dog meat is the least preferred meat from quality point of view in case of Konyak tribes (Table 2.a), Ao tribes (Table 3.a) and Angami tribes (Table 5.a). But preference of dog meat is in 3rd position in Lotha tribes (Table 4.a) from quality point of view and 4th in position from taste (4.b) point of view. In case of Konyak tribes scale value difference between 4th and 5th is 1.064 which indicate the degree of least preference of dog meat from quality point of view. Similarly it is also high in taste point of view and scale value is 0.771 and in case of other tribes there exist reasonable differences.

However, during winter period and Christmas occasion consumption rate is higher compared to other seasons. It is a good source of meat with so many traditional beliefs. As per respondents' view it has good medicinal properties. Traditionally people believe that it is good for generating body heat and body warmth. It increases the sexual potentiality like libido and virility. It is good for quick recovery from severe illness, and strongly recommended for lactating mother for quick rejuvenation of body condition.

Conclusion:

- Five popular sources of meat commonly eaten by the Naga tribes in Nagaland are pork, chicken, beef, mithun meat and dog meat.
- So far meat preference among major tribes of Nagaland, *viz.*, Konyak, Ao, Lotha and Angami are concerned, pork is the highly preferred meat followed by chicken, mithun, and beef. Dog meat is the least preferred meat among the tribes.
- Dog meat is gradually losing its status and acceptability.
- Price is not always related to preference of meat among poor tribes; whereas availability of meat in the local market is an important determining factor in this respect.
- Both quality and taste are the important considerations for consumption of meat among the tribes in Nagaland.

References:

Edwards, A.L. (1969) Techniques of Attitude Scale Construction. Vakils, Feffer and Simons Private Ltd., Mumbai.

Govt. of Nagaland (2006) 'Nagaland' Department of Tourism, Government of Nagaland, Kohima, Nagaland, India.

Raatan, T. (2004) Encyclopaedia of North East India, Kalpaz Publications, Delhi.

SASRD NAIP (2008) Annual report 2007-08, School of Agricultural Sciences and Rural Development, Nagaland University.

USDA (2006) USDA Foreign Agricultural Services Preliminary data 2006. www.fao.org.

* * *

Table 2.a: Z-matrix (Hierarchy of preference of meat quality by Konyak tribes)

Types of meat	Dog meat (C)	Beef (E)	Mithun meat (B)	Chicken (D)	Pork (A)
Dog meat (C)	0.000	1.341	1.341	1.645	1.126
Beef (E)	- 1.341	0.000	0.305	0.643	0.524
Mithun meat (B)	- 1.341	- 0.305	0.000	0.151	0.674
Chicken (D)	- 1.645	- 0.643	- 0.151	0.000	0.279
Pork (A)	- 1.126	- 0.524	-0.674	- 0.279	0.000
Sums-Z	- 5.453	- 0.131	0.821	2.160	2.603
Mean-Z	- 1.090 +1.090	- 0.026 1.090	0.164 1.090	0.432 1.090	0.520 1.090
RANK (Scale value)	0.000 5th	1.064 4th	1.254 3rd	1.522 2nd	1.610 1st

Table 2.b: Z-matrix (Hierarchy of preference of taste of meat by Konyak tribes)

Types of meat	Dog meat (C)	Beef (E)	Mithun meat (B)	Chicken (D)	Pork (A)
Dog meat (C)	0.000	0.994	1.175	1.405	0.878
Beef (E)	-0.994	0.000	0.202	0.583	0.806
Mithun meat (B)	-1.175	-0.202	0.000	0.000	0.878
Chicken (D)	-1.405	-0.583	0.000	0.000	0.176
Pork (A)	-0.878	-0.806	-0.878	-0.176	0.000
Sums-Z	-4.452	-0.597	0.499	1.812	2.738
Mean-Z	-0.890 +0.890	-0.119 +0.890	0.099 +0.890	0.362 +0.890	0.547 +0.890
RANK (Scale value)	0.000 5th	0.771 4th	0.989 3rd	1.252 2nd	1.437 1st

Table 3.a: Z-matrix (Hierarchy of preference of meat quality by Ao/Aou tribes)

Types of meat	Dog meat (C)	Beef (E)	Mithun meat (B)	Chicken (D)	Pork (A)
Dog meat (C)	0.000	0.300	0.445	0.659	0.776
Beef (E)	- 0.303	0.000	0.113	0.776	0.776
Mithun meat (B)	- 0.448	0.113	0.000	0.023	0.978
Chicken (D)	- 0.662	- 0.779	- 0.025	0.000	0.253
Pork (A)	- 0.779	- 0.779	- 0.982	- 0.253	0.000
Sums-Z	- 2.192	- 1.145	- 0.449	1.205	2.783
Mean-Z	- 0.438 0.438	- 0.229 0.438	- 0.089 0.438	0.241 0.438	0.556 0.438
RANK (Scale value)	0.000 5th	0.209 4th	0.349 3rd	0.679 2nd	0.994 1st

Table 3.b: Z-matrix (Hierarchy of preference of taste of meat by Ao/Aou tribes)

Types of meat	Dog meat (C)	Mithun (B)	Beef (E)	Chicken (D)	Pork (A)
Dog meat (C)	0.000	0.445	0.842	0.496	1.227
Mithun meat (B)	-0.448	0.000	0.300	0.445	1.454
Beef (E)	-0.842	-0.303	0.000	0.716	1.454
Chicken (D)	-0.499	-0.448	-0.719	0.000	0.842
Pork (A)	-1.232	-1.461	-1.461	-0.499	0.000
Sums-Z	-3.021	-1.767	-1.038	1.158	4.977
Mean-Z	-0.604 +0.604	-0.354 0.604	-0.207 0.604	0.231 0.604	0.995 0.604
RANK (Scale value)	0.000 5th	0.250 4th	0.397 3rd	0.835 2nd	1.599 1st

Table 4.a: Z-matrix (Hierarchy of preference of meat quality by Lotha tribes)

Types of meat	Mithun meat (B)	Beef (E)	Dog meat (C)	Chicken (D)	Pork (A)
Mithun meat (B)	0.000	0.189	0.454	0.935	1.150
Beef (E)	- 0.189	0.000	0.189	0.935	1.282
Dog meat (C)	- 0.454	- 0.189	0.000	0.524	0.842
Chicken (D)	- 0.935	- 0.935	- 0.524	0.000	0.253
Pork (A)	- 1.150	- 1.282	- 0.842	- 0.253	0.000
Sums-Z	- 2.728	- 2.217	- 0.723	2.141	3.527
Mean-Z	- 0.545 0.545	- 0.443 0.545	- 0.144 0.545	0.428 0.545	0.705 0.545
RANK (Scale value)	0.000 5th	0.102 4th	0.401 3rd	0.973 2nd	1.250 1st

Table 4.b: Z-matrix (Hierarchy of preference of taste of meat by Lotha tribes)

Types of meat	Mithun meat (B)	Dog (C)	Beef (E)	Chicken (D)	Pork (A)
Mithun meat (B)	0.000	0.189	0.674	1.282	1.645
Dog(C)	-0.189	0.000	0.126	0.935	1.440
Beef (E)	-0.674	-0.126	0.000	1.036	1.960
Chicken (D)	-1-282	-0.935	-1.036	0.000	0.385
Pork (A)	-1.645	-1.440	-1.960	-0.385	0.000
Sums-Z	-3.790	-2.832	-2.196	3.638	5.430
Mean-Z	-0.758 +0.758	-0.566 +0.758	-0.439 +0.758	0.727 +0.758	1.086 +0.758
RANK (Scale value)	0.000 5th	0.192 4th	0.319 3rd	1.485 2nd	1.844 1st

Table 5.a: Z-matrix (Hierarchy of preference of meat quality by Angami tribes)

Types of meat	Dog meat (C)	Mithun meat (B)	Beef (E)	Chicken (D)	Pork (A)
Dog meat (C)	0.000	0.000	0.253	0.755	0.842
Mithun meat (B)	0.000	0.000	0.063	0.690	1.036
Beef (E)	- 0.253	- 0.063	0.000	0.935	1.440
Chicken (D)	- 0.755	0.674	0.935	0.000	0.266
Pork (A)	- 0.842	- 1.036	- 1.440	- 0.235	0.000
Sums-Z	- 1.850	- 0.425	- 0.189	2.145	3.584
Mean-Z	- 0.370 0.370	- 0.085 0.370	- 0.037 0.370	0.429 0.370	0.716 0.370
RANK (Scale value)	0.000 5th	0.285 4th	0.333 3rd	0.799 2nd	1.086 1st

Table 5.b: Z-matrix (Hierarchy of preference of taste of meat by Angami tribes)

Types of meat	Dog meat (C)	Mithun meat (B)	Beef (E)	Chicken (D)	Pork (A)
Dog meat (C)	0.000	0.063	0.319	0.935	1.282
Mithun meat (B)	-0.063	0.000	0.063	0.690	1.150
Beef (E)	-0.319	-0.063	0.000	1.036	0.935
Chicken (D)	-0.935	-0.674	-1.036	0.000	0,454
Pork (A)	-1.282	-1.150	-0.935	-0.454	0.000
Sums-Z	-2.599	-1.824	-1.589	2.207	3.821
Mean-Z	-0.519 +0.519	-0.364 +0.519	-0.317 +0.519	0.441	0.764
RANK (Scale value)	0.000 5th	0.155 4th	0.202 3rd	0.960 2nd	1.283 1st

23

NEURO MUSCULAR DEFORMITIES IN LEAD INDUCED CHICK EMBRYOS-AN EXPERIMENTAL STUDY TO CORRELATE THE HAZARDS OF LEAD PRESENT IN ENVIRONMENT

S. K. PYNE * AND G SAMANTA **

Animal Science, Department of ASEPAN, Palli Siksha Bhavana, Visva-Bharati, Sriniketan, W.B.

ABSTRACT

Lead one of the toxic heavy metals, coming from various sources and ultimately reaches within the biological systems of animals. Here it has been tried to explore the effect of lead in neuro-muscular part of chick embryos. This study was made with 100 fertile White Leghorn eggs, divided into 5 groups. GROUP-I was kept as control, GROUP-II was injected with 1mg. lead nitrate/egg, and likewise GROUP-III with 2mg. lead nitrate/egg, GROUP-IV with 1mg, lead acetate/egg & GROUP-V with 2mg. lead acetate/ egg was given. In all the lead treated groups, only single dose of lead salt in 0.05 ml. distilled water was injected into the yolk sac on the 10th day of incubation.

Out of 80 embryonated eggs from treated groups (20 from each treated groups), more than 50% eggs failed to hatch. Eggs were opened and examined after schedule period (21 days) of hatching.

Distorted dead embryos were found within the egg shell. Among hatched chicks, 20% became lame within 2 weeks after hatching. However sensory nerves of the limbs were not involved for such disorders. In few cases wings were also affected (wing-drop). Affected chicks died within 21 days after hatching. It was noticed that both the lead compounds were detrimental to chick embryos; out of which lead acetate was more lethal than nitrate.

Introduction:

Lead is one of the toxic heavy metals. It is widely used in modern industries for its valuable properties. But lead as environmental pollutant can not be overstressed. The soil lead enters the ground water and ultimately reaches drinking water to be consumed by animals. The concentration of lead in air near heavy traffic may be twenty times more than that in rural areas. Uses of grass, bordering the roads carrying great deal of traffic for livestock feeding may result in deleterious consequences from nutritional point of view. Generally, heavy metals are known to have profound physiological effects even with trace quantities. Through various ways after entering into the biological systems how the lead affect neuromuscular system in the present experiment.

Materials and Methods:

Lead one of the toxic heavy metals, coming from various sources and ultimately reaches within the biological systems of animals. Here it has been tried to explore the effect of lead in neuro-muscular part of chick embryos. This study was made with 100 fertile White Leghorn eggs, divided into 5 groups. GROUP –I was kept as control. GROUP – II was injected with 1 mg. lead nitrate/egg, likewise GROUP – III with 2 mg. lead nitrate /egg GROUP – IV with 1 mg. lead acetate /egg & GROUP-V with 2 mg, lead acetate/egg was given. In all the lead treated groups, only single dose of

* Professor

** Professor and Head, Department of Animal Nutrition, WBUAFS Kolkata 37 lead salt in 0.05 ml. distilled water was injected into the yolk sac on the 10th day of incubation. Observations were made on 21st day of incubation during hatching, embryonated eggs failed to hatch and post-hatch period of 3 weeks. For this study 20 embryonatd eggs of each group was allowed to hatch.

Out of 80 embryonated eggs from treated group (20 from each treated groups), more than 50% eggs failed to hatch. Eggs were opened and examined after schedule period (21 days) of hatching. Distorted dead embryos were found within the egg shell. Among hatched chicks, 20% became lame within 2 weeks after hatching. However sensory nerves of the limbs were not involved for such disorders. In few cases wings were also affected (wing-drop). Affected chicks died within 21 days after hatching. It was noticed that both the lead compounds were detrimental to chick embryos; out of which lead acetate was more lethal than nitrate.

Results:

Out of 80 embryonated eggs from treated groups (20 from each treated group), above 50 percent eggs failed to hatch. Those eggs were opened and examined after schedule period (21 days) of hatching. In most cases, distorted dead embryos were fond within the egg shell. Among hatched chicks, 20 percent became lame (paralytic effect of legs) within 2nd week after hatching in case of treated groups. But in all lame cases, sensory nerves of the limbs wee not involved. In few cases wings were also became affected (wing-drop). Affected chicks died within 21 days after hatching. Rest of the chick including the control group were apparently healthy during the course of observations.

Discussions:

Failure of hatching in most cases (about 50%) and the distorted position of the embryos within the egg are most likely due to injuries of the brain and spinal cord resulting in the loss of behavioural reflexes and coordinated movements that the necessary for that complex process. Among the hatched embryos of the treated groups, most of them (80%)

were in good and healthy conditions. But in few cases (20%) the chicks became lame (paralysis of legs) without any sensory involvement. Oppenheim (1973) also observed the failure of lead-treated chick embryos to hatch within the scheduled period of hatching. Development of paraplegia and cerebellar damage of off springs were observed when lead acetate was induced to the sucking rat (Michael son and Sauerhoff, 1974). Silbergeld and Golberg (1974) however, concluded that lead inhibits the availability of choline for synthesis of acetylcholine in the synaptosome with increased levels of norepinephrine. Goyer and Rhyne(1973) and Hatch (1982) noted in mammals that peripheral lead neuropathy was characterized by segmental demyelination and axonal degeneration accompanied by Schwann cell degeneration. Neuromuscular syndrome, or lead palsy, muscle weakness, easy fatigue before actual paralysis with the involvement of extensor muscles due to lead toxicity were also described by Gilman *et al.*, 1980. They also noted that wrist drop and to a lesser extent food drop may be considered as pathognomonic for lead toxicity. Degenerative changes in motor neurons and their axons generally occurred without sensory involvement. Braunwald *et al.* (19897) however started that lead caused demyelization of prolong nerves, produced prolonged nerve conduction time and subsequent paralysis of extensor muscles (wrist drip of foot drops) with atrophy. Mortality of chick in the present study may be explained with the immunosuppressive effect of lead i.e. lead can suppress production of antibodies to certain viral and bacterial organisms, the mechanism of which is not known (Hatch, 1982).

Conclusion:

In the present investigation it is observed that both the lead compounds are detrimental to the embryonic tissues of advanced stages of development in chick embryos. Their effects are not always dose dependent. It is also found that the lead nitrate is more teratogenic than lethal effect and lead acetate is more lethal than teratogenic effect.

Bibliography:

Braunwald, E., Isselbacher, K.J. Petersdorf, R.C., Wilson, J.D., Martin, J.B. and Fauci, A.S. 1987. Harrison 's Principles of Internal Medicine. 11th edn. PP. 852-853, Mc Graw-Hill – Book Company, New York. Mexico, New Delhi, Tokyo, Toronto.

Goyer, R.A. and Rhyne, B.E. 1973. Pathological effects of lead. Int. Rev. Exp. Pathol. 12: 1-77.

Gilman, A.G., Goodman,. L.S. and Gilman, A. 1980. The pharmacological Basis of Therapeutics. 6th edn. PP. 1616-1622.Macmillan Publishing Co. New York.

Hatch, R.C. 1982. John's Veterinary Pharmacology and Therapeutics. Edited by Booth, N.H. and McDonald, L.E., 5th edn. Pp. 1005-1010. Kalyani Puboishers, Ludhiana , New Delhi .

Karnofsky, D.A. and Ridgway, L.P. 1952. Production of injury to the central nervous system of the chick embryo by lead salts. *J. Pharmacol. Expt. Therap*. 104: 176-186.

Michaelson, A. and Sauerhoff, M.W. 1974. An improved Model of lead induced brain dysfunction in the suckling rat. *Toxicol. And Appl. Pharm.* 28: 88-96.

Oppenhiem, R.W. 1973.Prehatching and hatching behaviour: Comparative and physiological consideration. Behavioural Embryology vol. 1, pp. 264-236. Academic Press, New York .

Silbergeld, E.K. and Goldberg, A.M. 1973. Lead induced behavioural dysfunction: An animal model of hyperactivity. *Exp. Neurol.* 42: 146-157.

* * *

24

INFORMATION COMMUNICATION TECHNOLOGY INTERVENTION FOR ENTREPRENEURSHIP IN RURAL DAIRY FARMING

RAM PRITI MANI TRIPATHI
Lecturer, B.R.D.P.G. College, Deoria (U.P.)

ABSTRACT

The potential using new ICT in India laid both intensive and small holder lives stock. The present paper will attempt to identify how ICT contribute for the continued growth and sustainability to this sector and focus on potential ICT application conventional, small holder based livestock production and services to livestock sector in India.

Introduction:

In the last four decade of twentieth century the face of agriculture turned from traditional to commercialization which got spurt with advent of green revolution in crop production; white revolution in the live stock farming towards milk production blue revolution in fisheries, yellow revolution in oil seeds, Rainbow revolution in bio-technological revolution, industrial revolution and information technology revolution. These revolutions enable the farmers to face the challenges of globalization of

Indian economy. With the globalization of Indian economy the challenges have become more intense in the field of agriculture with additional responsibility in food security in the twenty first century. The forecast in that by 2035 the population of India overtake China as the world's most populous country in the world out of the one billion 20 million population 390 million fall below poverty line in 2006. This obviously represents enormous human suffering; the children below the age of five years which amount of one third of Indian children are malnutrition. Therefore, in order to alleviate the poverty, agricultural growth is counted as main source. Thus, it is important to find new sources of production as growth of agriculture and requires recharging of agricultural development.

There is an urgent need of diversifying the agricultural sector to enhance the Indian economy. It is assumed that live stock production will become the key means of alleviating poverty in next twenty years. The poor will be benefitted through overcoming deficiencies in protein and micro- nutrients. Economic and social benefits have outpaced any other programme in rural development, women empowerment, food and nutritional security particularly among the nursing mother and children. The crop- livestock integrated farming system also provides an opportunities for promoting organic farming and restoration of soil health, and thereby of an evergreen revolution leading to an enhancement of productivity in perpetuity without associated ecological harm. It also strengthens household nutritional security. States which have hunger contribution from live stock in the state agricultural output and gross domestic product growth are the states which have les percentage of rural poor. State like Kerala, Haryana, Punjab, Himachal Pradesh, Rajasthan and Gujarat where the live stock share in agriculture output is a round 30 percent, rural poverty is only 10-15 percent, this is much below the national poverty figures. Therefore, challenges should be translated into opportunities. In changed rural and agricultural scenario of the country, entrepreneurship in agricultural development is now considered as the need of time. Women are a vital part of Indian economy constituting one third of national labour force and a major contributor to the survival of the family. The poorer the family, greater dependence on woman income. Despite progress in several key indicates, a gender analysis of most social and economic data demonstrates that women in India constitute to be relatively disadvantaged in matter of survival, health, Literacy, communication and productivity.

With the advent of science and technological development and emerging trend of society, women in rural areas have also started entering in to their professional roles, apart from the burden of working in home agriculture and allied activities. Their increasing desire of raised living standard and fulfillment of economic necessities also had initiated the entry of women in to various entrepreneurial activities. Entrepreneurship is neither science nor an art but creative and innovative response to the environment. Today's entrepreneurship has been recognized as the engine of economic change in particularly all parts of world. But to become a successful entrepreneur one requires a lot of knowledge, managerial competencies and hard work along with complete knowledge of entrepreneurship in a agricultural development is now considered as the need of time.

ICT base initiatives:

The emergence of information and communication technology (ICT) has brought revolutionary change in all field of human life. ICT provides access to latest updates of information on particular technology clearly thrust should lay greater emphasis on the transfer of scientific technological information from the research institute to its actual users. New ICTs includes community radio, Television Cellular telephone, use of computing device, digital imaging the internet and world wide area networking (WAN), Wi-Fi and mix media, for example use of radio for internet areas, Internet radio, SMS services and WAP (Wireless Access Protocol) base internet using cellular telephone as also embedded use of micro-processor. Computing devices and application of digital media in process and system for data collection and information management. It is well recognized that the computer images and video clips of actual application of agricultural technology can have greater impact on mind and psyche of rural, uneducated farmers than the textual description of technology with greater emphasis on establishing rural information Kiosks and Kiss an Call Centre along with country wide investment in creating information connectives ICT offers a more efficient attractive and interactive medium for information dissemination and provide novel opportunities for taking scientific knowledge/ technique to the end users.

The potential using new ICT in India laid both intensive and small holder lives stock. The present paper will attempt to identify how ICT contribute to the continued growth and sustainability to this sector and focus on potential ICT application conventional, small holder based livestock production and services to livestock sector in India.

There are three areas in which ICT can contribute to the livestock sector;

- Improved management of data and information
- Improved services including marketing.
- Improved learning capacity development.

Improved management of data and information

Accurate and precise livestock data and information is required for policy development strategy formulation, monitoring growth and evaluating impact of policies and programs for growth and sustainable development of the sector. The data and information is required at various levels such as:

➢ National
➢ Regional/ State
➢ District and Taluka
➢ Village, Community and Farm.

The purpose for managing this data and information can be for:

- Planning, implementation and monitoring livestock development program including those for animals improvement and effective natural resource management.
- Planning and monitoring livestock production at various level including village and farms.
- Planning and monitoring animal health, bio- safety and food safety.

- Planning and monitoring and capacity development.
- Livestock and livestock products marketing and supply chain management
- Planning livestock research

Data Collection and Management:

Numbers of information system have been developed to cater information needs in various areas that include crops, live stock production and agriculture marketing. The ultimate objective of managing livestock stock data and information would be related to making decisions for a given purpose at a given level. ICT can be used at each level starting from the village, handheld devices at the Computer and personal digital Assistants (PDA). Smart Card can be designed to inform the animal owner the expected next coming heat of the animal. The same system can be used to plan rounds of the inseminator for information making a passive information system into active one and enabling monitoring of fertility at the animal level and through aggregation at various other levels such as village, Taluka, inseminator even the sire. The potential benefits of introducing an information system around low cost computing devices are immense when the cost of a missed heat for insemination or infertility can be used as 1000 rupees a month.

In order to understand the issues of using ICT Can have combination of multipurpose. Apparently, along with the use of ICT, there will be a vital need to manage, coordinate and integrate information system in various level and for various uses. In India, the use of personal computers by individual small holders may not be possible in the near future. However with lowering of costs and introduction multipurpose information Kiosks at community and village level such as health, education, agricultural extension and governance or internet Cafes and even with the retail shops, the use smart cards and store the farmers and animal owner information can be introduced. An advantage of creating such a system would also enable to animal owners. Providing a title enable the farmers to monitor his/her live stock and use of this also for financial transactions especially as mortgage of loan or an advance. This can have a huge impact on poverty alleviation of live stock owners who are usually poorest of the poor.

The potential economic contribution of prices and accurate data and information can be enormous. The ICT for enabling these information systems is now commonplace. It is becoming cheaper and can be very cheap if used for masses. These information systems will be need greater coordination and management as also development of public data base, decision support and knowledge base tools. Compared to live stock sector contribution to India's economy investment in these systems will be miniscule and as live stock sector get more organized, the access of these information systems can be commercialized. What is really needed is greater awareness and sensitization among the policy makers about the potential of benefits of ICT's and improved information system in contributing to the development of the live stock sector.

The most obvious application of new ICT in veterinary services in India is the use of cell phone. The cell phone enables the veterinarian to plan provide services and maintain animal records. This form of veterinary telemedicine that has not attempted to exploit the ability to collect, collect and integrate data through information system enable the provision of preventing veterinary medicine, herd and folk health at the village level. This is the direction that public veterinary services will need to take the benefit for small holders.

Improved Market Intelligence:

As agriculture and with it live stock production becomes more market oriented in India. The information in commodity chain therefore also a very short cycle time. This makes use of ICT to manage this information obvious. Easier availability of market information including prices and trends has potential to charge the structure of the present live stock production system in India. Number of information system have been developed to cater information needs in various areas that includes crops, live stock production, fisheries, agricultural marketing and information network (AGMARKNET) is meeting the information needs of the farmers by providing daily prices of agricultural prices on commodity in various man dies for helping the farmers to decide where to sell his produce.

Capacity and skill development of live stock producers:

At the movement, there is very little work in India to use ICT to communicate information and create learning among members of this innovation system. To exploit potential of live stock production by small holder in India and complete in a global market will require huge effort in capacity building development. The capacity building development need not only among the producers but also across the entire live stock commodity chain which would also include also service providers, processors, market intermediaries, and consumers etc. ICT to communicate information and create learning among the members of this innovation system. Under the SEWA provides women training in Gujarat with basic computer education to help them manage micro enterprises. The National Dairy Development Board (NDDB) in Gujarat is digital milk collection and thereby helping farmers. Beside these information system portal sites on agriculture accessible on internet providing a large amount of agribusiness related information covering more than 15 sub-sectors within the agriculture and food industry. This portal also provides useful information related to the commodity prices at national and international market news analysis buy and sell trade leads, Government policies.

ICT Kiosks for use by the farmers providing information on production, processing and marketing. For enhancing live stock productivity and to make communication more effective amongst the rural women. Agriculture and Food industry viz-rice, wheat, spices, sugar, cotton, fruits and vegetables herbs and medical plants dairy and poultry etc. These portals also provide useful information relating to the commodity prices at national and international markets news analysis buy and sell trade leads Government policies.

For enhancing live stock productivity and to make communication more effective amongst the rural women. The study was conducted in three blocks of Gorakhpur district and data were collected from 120 women with the help of structured interviews schedule. The study revealed that

50% of the respondents belonged to middle age groups and had nuclear family system with a low level of literacy (75%). Agriculture was the main occupation of about two-third of the respondents in both exposed and unexposed villages. Majority of the respondent have size of land holding of less than 2 Hectares. Small size herd was maintained by 82.6% of respondent in both set of villages. About 67% of respondents of unexposed and 45% in exposed villages have favorable attitude towards training of dairy farming. Hierarchically, pleasure motives secured first place followed by scientific and affiliation motives in performing the animal husbandry tasks. About 58% of women perceived medium level of training needs. In fixing priority areas of training in animal husbandry activities, health care appeared in first place follow by feeding, breeding and management.

For better technology transfer and effective Information Communication Technology majority of women preferred summer season followed by winter and rainy season, short duration training (15 days) in their own villages preferable by middle aged communicators (60%). Respondent opted to have medium size of homogenous groups (10-20) with respect to sex (80.8%) villages (70.8%) and socio-economic status (50%) for better and deep interaction during entire courses of learning. About 46, 31 and 23 percent respondents favored skill knowledge-cum-skill and knowledge oriented training for maximum learning of technologies related to animal husbandry. Film was one of the most liked communication tools in learning situation revealed by 60% of the respondents followed by film strips, models and specimens in descending order for sustainable learning. The lecture was preferred by 80% of the respondents as a basic and initiating communication tool of technological learning followed by group discussion, demonstration and field visits.

Changed scenario calls for incorporation of these perspective information communication technology. Till now improved farm technology were handed over to women farmers become more productive in terms of higher farm and dairy produce. No doubt ICT created a greater awareness among rural women thus contributing a lot to our "food self-sufficiency status" of now a day. Role of ICT is not only to transmit information regarding modern farm technologies but also to promote rural entrepreneur ship for overall agricultural development.

References:

ICAR 2007-08 Annual Report 2008 Indian Council of Agricultural Research.

I.I.E.P. (2007) Expending Training Modern Computer based research methodology retrieved from-

http/www.unesorg/iiep/eng/research/edual/comompares.htm.

Adhiquri, P (2006) "Research and Extension linkages in Indian Agriculture" country paper on SAIC Regional Workshop on research extension linkages for effective delivery of Agricultural technologies in SAARK countries 20-22 November 2006 NAARM Hyderabad, India.

Rice M.F. 2005 information and communication technologies and global digital divide technology transfer, development and least developing countries cooperative technology transfer and society vol.I. No.-1 PP72-87.

Briz-Kishore B.H.(2004) 'A focused Strategy Towards Entrepreneurship and Employment for realizing the knowledge connectivity in PURA University News 42(35) Aug-Sept P40-524.

Gupta M and Mandope M.K.(2003) communication and technological empowerment of rural women in dairy farming paper presented, International conference on communication for development in information Age: extending benefits of technology for all at Varanasi B.H.U. with association of Global Communication Research Association.

COI (2001) Census of India 2001.

Bathrick David D (1998) "Forecasting Global well being A-New Paradigm to Revitalize Agricultural and Rural development" 2020 Vision International Food Policy Research Institute 2033 N.W.Washington D.C. 20006 U.S.A. Sep. 1998.

M.S.Swaminathan (1992) Servey of Indian Agriculture P-6, preparing for live stock production "World Food Prospects" critical issues for the early twenty first century Pin Pinstrips Anderson Raju Pandaya-lorch Mark W. Rosegrant Food Policy Research Institute 2020 Vision P-3.

* * *

25

AN APPROACH FOR IMPROVING LIVELIHOOD SECURITY THROUGH SUSTAINABLE PIG PRODUCTION IN NEH REGION

S. Naskar[1] and Nilotpal Ghosh[2]

[1]Sr. Scientist, Indian Veterinary Research institute,
Eastern Regional Station, 37, Belgachia Road, Kolkata -700037
E-mail: s_naskar@rediffmail.com

[2]Associate Professor, Department of Animal Science,
Bidhan Chandra Krishi Viswavidyalaya, Mohanpur,
Nadia, West Bengal – 741252
E-mail: gnilotpal@yahoo.com

ABSTRACT

The importance of livestock production, particularly pig husbandry has given special emphasis in the north-eastern region of India, as it is the most important livestock and pork being the preferred meat item among the non-vegetarian folk of people in this area. This region has a sizeable population of pig which constitutes around 28.10% of the country's total pig population, and almost every rural household of Nagaland, Mizoram and Meghalaya keeps at least few pigs as a subsidiary source of income. Many exotic pig breeds are experienced in this region like *Large White Yorkshire, Hampshire, Landrace, Duroc, Tamworth* and *Large Black,* though majority of the pig population of this area constitutes non-descript indigenous germplasm. The motto of the pig production is to help rural peoples who are economically and socially weaker section of the society. Lack of superior germplasm, lack of slaughter house, non-availability of feed, more disease incidence and lack of knowledge about scientific package of practices for improved pig husbandry are some of the constraints of pig production in this area. This article focuses these aspects which should be taken in to consideration for improving the pig production in this part of the country.

Keywords: Livelihood security, pig, north-east hill region

Introduction:

Livestock Production is not only an integral and inseparable part of Indian Agriculture and Northeastern states in particular. It plays a vital role in the socio-economic condition of the farmers. Percentage of rural population in NEH region ranges from 85% in Meghalaya to 96% in Arunachal Pradesh, whereas the all India average is 80%. Most of these rural people are doing agriculture by occupation. A major portion of the farmers are basically small and marginal farmers. The total land availability for agriculture in this region is 2270 million hectors and out of which 70% is located in Assam, 8% in Manipur and 7.5% in Tripura. The area available for other states is very negligible. Therefore, farmers find animal husbandry to be a better alternative. Moreover, the tribal people of NEH states are exclusively non-vegetarian in their dietary habits and they prefer meat and meat products. Tribal people constitute 79% of the population in Arunachal Pradesh, 88.6% in Nagaland and 80% in Meghalaya. There is no taboo of taking any type of meat and their most preferred meat is pork. Therefore, importance of livestock production, particularly the piggery has given special emphasis in this region. Several indigenous and exotic breeds of pigs are available throughout India.

An Overview of Pig Production in NEH Region

Rearing of pig is one of the important animal husbandry practices by the down trodden tribal people of India. It is mainly managed by nomadic system of rearing with the weaker sections of society as a source of income as well as for family consumption. Cost of rearing is very much negligible and return is not also a matter for them. A bulk of pig population in India is indigenous type with very low productivity. Similar scenario is also prevailing in northeastern states. Pig is the most important livestock in NEH Region. Almost every rural household of Nagaland, Mizoram, Meghalaya keeps at lest few pigs as a subsidiary sources of income. The region has a sizeable population of pig (Table 1), which constitutes around 28.10% of the counties 13.58 million pigs. Rearing of pig in India as a whole and this region in particular is still in poor condition, because certain rural people who are economically and socially weaker section of the society mostly raise pigs.

Table 1: Pig population in Northeastern states of India

(in thousands)

States	1982	1992	2002	Decadal Change (%)		Annual growth (%)	
				1982-92	1992-2002	1982-92	1992-02
Arunachal Pradesh	213	239	330	0.12	38.08	0.01	3.81
Assam	514	1364	1543	165.37	13.12	16.54	1.31
Manipur	176	383	415	117.61	8.36	11.76	0.84
Meghalaya	207	294	419	42.03	42.52	4.20	4.25
Mizoram	77	112	218	45.45	94.64	4.55	9.46
Nagaland	248	526	644	112.09	22.43	11.21	2.24
Sikkim	30	45	38	50.00	10.00	5.00	1.00
Tripura	103	188	209	82.52	11.17	8.25	1.12
NE States	1568	3151	3816	100.96	21.10	10.09	2.11
All India	8499	12795	13578	50.55	6.12	5.06	0.61

Production potentialities of pigs:

The pigs are probably the most accommodative farm animals. Pigs can be reared economically with minimum expenditure. The generation interval of pig is short and prolific in nature. The stock can be increased or decreased to the desired number at any time. With a large stock, there is greater scope for proper selection for fattening and breeding purposes. The quantity of meat available per unit live weight of pig is larger than other kinds of livestock. Pork also contains more energy than other meats per unit weight. Pig husbandry requires minimum labour and returns over investments are quicker. Pigs utilize garbage, garden waste and discarded feeds very effectively. They are the most efficient converters of feeds into edible meat. Considering the acute paucity of protein supply and the increasing demand for animal protein on the one hand and the high production potentials of a well organized pig industry on the other, it is evident that the improvement of pig production in the poorer parts of our country has greater promises.

Different pig breeds in NEH Region:

Several indigenous and exotic breeds of pigs are available throughout India. Most commonly available exotic pig breeds are Large White Yorkshire, Hampshire, Landrace, Duroc, Tamworth and Large Black. These breeds are comparatively heavier than local breeds. Their growth rate, prolificacy and litter traits are better those local pigs. Indigenous pigs can survive under stressful environmental condition like poor nutrition, high environmental temperature, high rainfall and humidity. Local pigs can be reared with very low inputs. This makes the traditional pig industry profitable in India as a whole and northeastern region in particular. Basically the names of the pigs breed originates either on tribe / area / state like Doom, Mali, Mizo / Tuikuk, Naga local, Meghalaya local/khasi local and Angamali breed. These pigs are very small in size as compare to exotic breed, potbellied, hairy and mostly black in colour with elongated snout and erected ears

System of Pig Production:

There are three main systems of pig production in use throughout the world. These are scavenging system, backyard system and commercial system.

1. Scavenging system

Subsistence farmers in many parts of the country practice this. Under this system, the owner allows his pigs to roam and find their own food, which consists of grass, roots, nuts and insects from soil. When natural food is scarce, supplementary food may be given. In towns, villages, the pigs are allowed sometimes, to root around in refuse dumps. Production by this system is very low. The pigs are seldom sold but are kept for use by the farmer's family and sometimes for community.

2. Backyard system

In this system, a small numbers of pigs are kept in a backyard. A variety of cheap housing methods are generally used. The pigs are fed on kitchen waste materials supplemented by cereals and by products. A sow may be kept and some or all, of her offspring sold as soon as they weaned. Sometimes, one or more piglets are purchased and fattened for family as well as for sale.

3. Commercial system

For commercial pig producer, this enterprise is a major source of income. The use of efficient production techniques is important. The pigs kept are generally improved breeds with a potential for high productivity. The management plays an important role in pig farming. The management of the herd affords the greatest opportunity for increasing profit in swine production. Reproduction traits are very low in habitability and, therefore, more dependent upon the management than the genetic make up for efficient performance for the breeding herd. The management includes daily routine handling the animals within the limitation of the conditions. Planning and modification will be increasing the effectiveness of the management.

Major Constraints in Pig Production in NEH Region:

1. Lack of superior germplasm for higher productivity.
2. Non-availability of pig feed.
3. More disease occurrence of various diseases due to heavy rainfall and high humidity.
4. Complete lack of organized slaughterhouse on this region for scientific slaughter and hygienic packaging of pork and pork products for proper marketing through out the region.

Lack of Superior Germplasm:

Pig population of NE state is 3.816 million that contribute 28.10% of the country's pig population. Indigenous (desi) pig population still occupies more than 56.68 % of the total population. Decadal change in desi pig population (Table 2) indicated that the population is changing gradually even after implementation of improvement programmes. If we calculate the 100 of pigs per 100 human, the figure is much higher than national figure. Even then, the region has been importing pigs from outside this region to fulfill the demand of this region because of high rate of pork consumption, low production of indigenous pigs, which shares a huge portion of total population, Lack of systemic breeding plan to improve the productivity of indigenous pigs and traditional system of pig rearing.

Pork consumption is higher in this region as compared to other parts of the country (Table 3). This may be due to preferences of pork by the tribal peoples of NE states.

Table 2: Decadal change of indigenous pig population

States	1982				1992				2003			
	Exotic & C.B	***Desi***	**Total**	**%** ***Desi***	**Exotic & C.B**	***Desi***	**Total**	**%** ***Desi***	**Exotic & C.B**	***Desi***	**Total**	**%** ***Desi***
Arunachal Pradesh	21	192	213	90.14	9	230	239	96.23	8	322	330	97.58
Assam	72	442	514	85.99	299	1065	1364	78.08	491	1052	1543	68.18
Manipur??	40	128	176	72.73	132	251	383	65.54	208	207	415	49.88
Meghalaya	31	176	207	85.02	102	192	294	65.31	188	231	419	55.13
Mizoram	42	35	77	45.45	73	39	112	34.82	196	22	218	10.09
Nagaland	56	192	248	77.42	244	282	526	53.61	361	283	644	43.94
Sikkim	6	24	30	80.00	9	36	45	80.00	19	19	38	50.00
Tripura	5	98	103	95.15	12	176	188	93.62	95	114	209	54.55

Table 3: Meat production and requirement (million tones per year) in NE states

State	Production	Projected requirement	Deficit %
Arunachal Pradesh	0.010	0.012	16.7
Assam	0.127	0.293	43.3
Manipur	0.019	0.026	26.9
Meghalaya	0.020	0.025	20.0
Mizoram	0.012	0.010	16.7
Nagaland	0.020	0.022	9.1
Sikkim	0.0041	0.006	31.7
Tripura	0.004	0.035	88.6
Total	0.216	0.429	49.7

(Meat requirement was calculated as 11 kg / capita / year)

Table 4: Availability of egg, meat, fish and milk (per capita) per year in NE states

State	Egg (Nos)	Meat (g)	Fish (g)	Milk (ml)
Arunachal Pradesh	32.0	17.05	2029	41.24
Assam	18.3	4.77	6001	27.52
Maninpur	29.7	7.83	6.72	28.05
Meghalaya	36.7	13.40	2.68	26.89
Mizoram	48.2	12.91	3.21	20.02
Nagaland	25.6	10.06	2.77	24.64
Sikkim	44.6	7.59	0.09	68.52
Tripura	32.9	1.28	9.22	15.35
Average	33.5	9.36	4.12	31.53

Poor Productivity:

Productivity and reproduction performances of local pigs, exotic and crossbred pigs were studied were at Institute as well as village level and result revealed that production performances of local pigs are significantly than crossbred as well as exotic pigs.

Lack of Systemic Breeding Programme:

A significant number of pig breeding farms were established to improve the indigenous pigs for quantum jump of pig production. Various programmes were started for improvement of indigenous pig through upgrading with exotic pigs but expected result has not yet been achieved due to lack of systematic breeding plans. These may be due to lack of proper implementation of breeding plan, insufficient infrastructure and skilled manpower. The sporadic attempt has led to a mixture of crossbred pigs without a fixed percent of exotic inheritance. This approach has to be stream lined and worked on identification, preservation and multiplication of one or two breeds in scientific manner.

Insufficient Numbers of Breeder Farmers:

Most of the pig rearers are small and marginal farmers and have very mere resources to invest on piggery. Pig breeding farms requires some capital investment at initial stages for infrastructure facilities like housing, base stocks and feeds. But most of these small and marginal farmers do not have the capacity to invest such amount. They always prefer to rear one or two pigs for fattening purpose. In this system, farmers sales their stock and look for a new stock every time which is very limited. So, the State Govt. should encourage the progressive farmers to be breeder farmers by helping them desirable package of pig production technologies as per the requirement and financial assistance.

Other problems are:

- Pig production pattern in NEH states is completely different from developed country i.e. production by masses compared to mass production by a few.
- Pig rearing is scattered among the landless labourers, small and marginal farmers and tribal masses.

- The people involved in piggery sector are mostly illiterate and lack of knowledge of proper feeding and improved management practices.
- The economic condition of pig growers are so poor that they are unable to adopt newer technologies maintain improved breed and purchase the modern tools.
- Practically non-existence of sufficient number of abattoirs and pork processing plants production of hygienic pork and pork products.
- Lack of organized marketing infrastructure and channel.

Strategic Issues:

In order to achieve the goal, time has come to address the issues like,

1. Develop a more effective understanding of the available pig genetic reason and their conservation and management.
2. Determine the available feed resource and their utilization, identification of non – conventional feed and also to improve nutritional quality of existing feed resources through mechanical chemical and microbial treatment.
3. Develop suitable disease surveillance, monitoring and forecasting system and develop control measure.
4. Develop suitable methods for efficient handling, processing and storage of pork products in order to reduce loss and increase self-life.
5. Use modern biotechnological methods for resources conservation, genetic improvement through reproductive practices.

Approaches to Achieve the Goal:

Indigenous pigs of NEH Region are not suitable for commercial farming due to its poor growth rate and feed conversion efficiency as compare to exotic /crossbred pigs. In rural areas still farmers prefer local pigs because they survive under very poor feeding and management condition with nominal input cost. Study clearly indicated that the crossbred developed between local pigs with exotic Hampshire, performed better both in terms of growth rate and prolificacy with increasing inheritance. Similarly, purebreds maintained under standard feeding and management was also found to perform better. So to make the region self sufficient in pork, following policies for improvement of pig production has to be taken under the collaborative manner among ICAR, other central govt. bodies, state government and NGOs.

Breeding Policies:

1. Encouragement of commercial farming of improved exotic purebred pigs with proven potential. Few elite populations of breeds like Hampshire; Yorkshire etc. are maintained for providing quality germplasm to the upcoming entrepreneurs.
2. Attempt to boost commercial farming of crossbred of different breed and also crosses of exotic with indigenous may also be made in the same way.
3. Upgrading of indigenous pigs by use of exotic boar in rural areas may be done keeping the interest of poor farmers in mind.
4. Measures have to be taken to conserve the indigenous pig germplasm and prevent then from unwanted breeding. Improvement of these indigenous pigs should be done through selective breeding.
5. Encourage the artificial insemination method for quick genetic improvement of indigenous pigs and get rid of the problem of breeding boars.

Alternative Feed Resources:

Feed cost shares roughly 80 percent of the total recurring expenditure. This region has already been running in shortage of food grains. Pig is the competitor of human food. So, emphasis should be given to non-conventional feed resources. Simultaneously, encourage to farmers to cultivate more food grains.

Infrastructure Development:

Here is an urgent need to boost up the infrastructure facilities of pig breeding farms, which has already been working in different parts of all the Northeastern states. They should continue with well-defined policies in respect to breeding and keeping the demand of the area.

Modern Disease Diagnostic Laboratory:

This region should have modern disease diagnostic laboratory followed by branches of each and every states to keep the stick vigilance of diseases and prevention of it. There should be well-planned network to work with different state veterinary departments.

Slaughter House:

This region also needs a well equip modern slaughterhouse for making pork and pork products. Still pigs are slaughtered in traditional methods with maintaining any hygiene. At least each state should have one modern slaughterhouse.

References:

Anonymous (2002). Basic Statistics of NEH. NEC, Shillong.

Bujarbaruah, K.M. and Das, A. (2002). Present status and future conservation and management of pig genetic resource. Key paper presented at X International Congress of the AAAP, 23rd to 27th September, 2002. New Delhi.

Das, Dharmeswar (2001). Pig production potential in Assam. In: Agriculture in Assam, Assam Agriculture University.

Sakurkar, C.S. (2000). Piggery India Year Book, 2000.

* * *

26

INDIAN WOMEN IN INDIAN DAIRY BUSINESS-A CASE STUDY OF CHITOOR DISTRICT OF ANDHRA PRADESH

Ashish.K.Makwana
Assistant Professor, Dairy Business Management Department College of Dairy Science Anand Agricultural UniversityAnand-388110, Gujarat
E-mail: ashish_gau@rediffmail.com

Introduction:

Comparing man and women on different indices of development and socio economic status, women are lagging behind men on all fronts throughout the world. It starts from cultural construct of son preference leading to the psychological difference to female child. In work, employment, earnings, education, health status and decision making powers, there is a clear difference between men & women. Women constitute fifty percent of the world's population and produce fifty percent of food across the globe but receive only ten percent of income. Women's access to and ownership of resources is less than men and even among those who own property, the control of its use and dispensation vests more often with any male member of the house. In many communities women cannot or do not inherit land. Thus it is difficult for them to earn through cultivation. Women's average earnings are consistently lower than men. In unorganized sector their earnings are even lower. With regards to education, literacy figures are lower for females compared to males. Political arena is also to a great extent monopolized by men. Several studies have recorded a greater gender asymmetry in the utilization of health service during illness. Fewer women than men seek and receive treatment. Violence against women has its roots in the subordination of women at the social level. Empowerment of women in such a scenario calls for modalities other than mere economic betterment.

Empowerment means different things to different people in different situations and countries. Empowerment is a process that helps & assist women realize their identity, capacity, capability, strengths & power & its application in day to day life in various spheres. It enable them to have access to resources, higher say in decision making, more ability to plan their time & family size & also free them from irrelevant customs, traditions, practices & prejudices. Empowerment of women thus lies in transcending from a lower level to higher level of welfare, access, participation and control with dignity. Women's empowerment in this context involves women recognizing, first of all their existing economic role, whether waged or unwaged, secondly enabling them to organize to gain social and even political reorganization of this role. Finally a package of skill enhancement, capacity building, credit facilities, educational inputs (literacy, knowledge of laws and rights, vocational training) and bargaining capacities raising women's economic position at par with or better than that of man of the same group. This requires organizing women through some kind of economic interaction programme, providing credit facilities, giving training, etc. and making them conscious about their economic position & power. Dairying provides one such avenue.

Research Findings:

A study was conducted by following ex-post facto research design and random sampling technique in Chitoor district of Andhra Pradesh by B Subrahamanyeswari and others to know about entrepreneurial behaviour of farm women.. A total of 120 respondents comprising 61 small, 35 medium, and 24 large rural women farmers in dairy were selected by proportional random sampling techniques. The results revealed that majority of the dairy women farmers possessed medium entrepreneurial behaviour and variation among the three categories of the respondents was found to be significant ($p<0.01$), management orientation ($P<0.01$), value orientation ($P<0.01$), educational ($p<0.05$) and innovativeness ($P<0.05$). Most of the variation in entrepreneurial behaviour was explained by the variable namely dairy farming experience, social participation, training received and income from dairy farming.

A study was conducted in Thrissur taluk of Thrissur district to assess the training needs of farm women engaged in dairy farming. It was found that out of the five major farm operations studied, the farm women needed training the most in the highest order was in housing. The minor operations preferred the most for knowledge need were proper design of cattle shed, selection of breeds, compounding balanced feed using locally available ingredients, vaccination and banking and insurance. As for skill need, selection of breeds, compounding balanced feed using locally available ingredients symptoms of common disease and balancing and insurance were preferred the most.

A study was undertaken to study the strategy followed by 'Swapnabhumi',a non government organization for promoting dairy of Women Self Help Group (SHG) group in Gangakhed taluk of Parbhani district. The objective of the study was to know the details of the dairy managed by women SHG under swapna bhoomi kerwadi (NGO). Increased self esteem, leadership development, risk bearing ability and decision making power among rural women were the perceived gains of the project. The amount from dairy was paid to women; thereby money was in the hands of women. In addition the ownership of buffaloes was given to women only. Because of this, the women were empowered and increase in their self confidence was noticed. Due to consumption of milk daily, the nutrition status of family members particularly of children was found better.

A study was conducted to analyze the impact of micro finance on income and employment of rural women in dairy enterprise. The study was conducted in tank commands of Haveri and Bellary districts where Community Based Tank Management Consultancy Project (CBTMCP),

University of Agricultural Sciences, Dharwad extended micro finance to poorest among the poor women to take up dairy enterprise as Income Generating Activity (IGA) on pilot basis. Data collected through pre–tested schedule by personal interview method during 2007-08. The study revealed that majority of the beneficiaries were middle aged (50%), illiterate (50%), married (93.33%) and belong to nuclear family (90%). The total maintenance cost per annum per buffalo was found Rs. 9937.23 in Haveri district and Rs. 10306.17 in Bellary district. The net return was found to be Rs 3945.56 and Rs 4581.06 per annum in Haveri and Bellary districts respectively.

Role of women in Dairying:

In the villages, the women are involved in agricultural & dairy husbandry's work as a family member. Consequently they work as a wage earner to their own farm as labour by doing various farm operations. As compared to other sectors, the participation rate of women in dairying sector is more than men. It is 65% as against 31% for men. Women have always played a major role in the dairy sector, looking after the cattle, caring and feeding them and maintaining the herd. Depending upon the economic status, women perform the tasks of collecting fodder, collecting and processing dung. Dung composting and carrying it to the fields is undertaken by women. Women also prepare cooking fuel by mixing dung with twigs and crop residues. In animal husbandry, women have multiple roles. They perform activities such as care of animals, grazing, and fodder collection, cleaning of animal sheds to processing milk and livestock products. In livestock management, indoor jobs like milking, feeding, cleaning etc are done by women in most of the families.

Inhibiting factors for women in Dairying:

Even though women's role in livestock management is significant, control over livestock and its products is very less. The income from dairy animals does not remain in their hands and they also do not have significant say in decision regarding sale and purchase. The constraints affecting rural women's ability to improve yield, profit and efficiency in animal husbandry and dairying activities include

(a) Illiteracy on the part of women is responsible for lack knowledge and understanding of basic concept of performing various farm operations.
(b) Gender based work differentiation leads to excessive work burden of both house management, farm operations, and child rearing on women.
(c) Social concept of male dominated society.
(d) Opposing view about social & constitutional laws about inheritance of property in India

Government interventions to empower women in Dairy activities

Looking to the present position of women in agriculture/dairying and their limited access to research & extension service it is necessary to adopt entirely innovative approach towards woman farmers to empower them so as to cater to their existing needs & emphasizing the future challenges.

India is largest milk producing country in the world but the bacteriological quality of milk is inferior compared to developed nations. This acts as a major hurdle for exports. Making women aware about importance of better hygienic practices and improved animal husbandry practices through training programmes is the first and foremost requirement.

Realizing the law status of women, Government of India has taken number of interventions for their socio economic improvement. Support to training and employment to women (STEP) is one among them. Department of Women and Child Development, Ministry of Human Resource Development, Government of India is implementing the scheme. Women Dairy Project (WDP) is one of the components of STEP. It is a holistic programme in which various sub systems are adequately looked into so as to achieve system's overall objectives. Under WDP, women dairy cooperative societies are only promoted. These societies are owned, controlled and managed exclusively by women. Apart from providing financial assistance for planning and management of these societies, 10-15 % of the budget of the scheme is devoted for training.

National Dairy Development Board (N.D.D.B) has adopted suitable polices for improving status of women engaged in dairy activities. With the efforts of N.D.D.B, today India's organized cooperative dairy network as on march 2009 includes 177 unions, operates in 346 districts, covers 1,33,349 village level societies and is owned by around 13.9 million farmers members of which 3.9 million are women.

NDDB has launched The Women's Dairy Cooperative Leadership Programme (WDLCP) as a pilot programme with the objective of strengthening the dairy cooperative movement by significantly increasing women's participation as active members and as leaders in the governance of cooperative societies, unions and federations. NDDB provides assistance to milk producer's cooperative unions for conducting various activities to achieve WDCLP objectives. The activities taken up under WDCLP include:

1. Awareness generation: (a) Health and Sanitation: Arranging health and sanitation camps, Campaigns and education programmes for women focusing on material and child health and disease prevention. (b) Improved animal husbandry practices: Demonstrations, field visits and other programmes to inform women of appropriate feeding, breeding, calf rearing and milch animal management practices.
2. Socio economic activities: Organization of mahila mandals (Women's clubs), Women's literacy programmes, Organization of thrift and credit groups
3. Leadership Development: Training programmes and exposure visits to potential leaders, Training of village resource persons.
4. Cooperative governance: Education programmes on Dairy Cooperative society activities, member responsibilities and rights, managing committee member duties

Conclusion:

In India dairying activity is undertaken as a small holder activity. In other sectors, Indian government is committed to adopt 33 percent reservation for them. But in dairy sector, majority of activities are performed by them. It is necessary to give them due credit for their efforts. This paper would be incomplete without highlighting the contribution of Dr Miss Amrita Patel, recipient of Padma Bhusan and many more national and international awards and at present acting as chairperson of N.D.D.B. If India is the largest producer of milk in the world today with 104.8 mt. of production in 2009, a good share of the credit also goes to Dr Amrita Patel, along with Dr Varghese Kurien, the Father of India's White Revolution. She was part of the team that conceptualized and more importantly, implemented 'Operation Flood' - a national dairy development programme - that helped promote, finance and deliver a variety of supporting services to a national cooperative dairy structure. The programme contributed significantly in increasing the per capita availability of milk in India and also increasing producer's income in both relative and real terms.

Apart from increasing milk production, Dr Amrita Patel placed a major emphasis on women's education as part of the co-operative development programme. Men were educated about the role of women in dairying while women were motivated to join dairy co-operatives and assert their rights as member. They were also encouraged to stand for membership of managing committees and in some states co-operatives and unions began to reserve seats on their boards for women. Today, more than 6,000 dairy co-operative societies in India are women's societies. The percentage of women members has risen to more than 20%. There have been small increases in the numbers of women serving on co-operative managing committees as well as in paid positions.

References:

Antwal, P.N et. al (2009) "Women in dairy management: a case study" *Journal of dairying foods and H.S*, 28(1) pp 64-66

Durga rani V and subhadra M.R (2009) "Training needs of farm women in dairy farming" *Veterinary world*, 2(6) pp 221-223

Grover Indu and Grover Deepak (2002) "Empowerment of women" Geeta somani Agrotech Publishing Academy, Udaipur-313002

K, Suresh et. al. (2009) "Micro Finance and empowerment of rural women: A case study of Dairy enterprise" *Karnataka Journal of Agricultural Science* 22 (4) pp 831-834

N, Akila. R et. al. (2007) "Women empowerment in Animal Husbandry enterprises" *Indian Dairymen* 59 (6) pp 57-60

Narasimhan Sakuntula (2009) "Empowering women: An alternative strategy from rural India" Sage publication, New Delhi-110048

Rao, V.M (2005) "Cooperative and Dairy development: Changing destiny of rural Women, Vedams Books.

Subrahmanyeswari et al (2007) "Entrepreneurial behaviour of rural women farmers in dairying: A multidimensional analysis" *Livestock research for rural development* 19(1) pp 23-25

www.nddb.org

* * *

27

A- REVIEW ON STATUS OF DUCK DEVELOPMENT IN MANIPUR

K. Rashbery Singh
Deputy Director (Instruction), Central Agricultural University,
Iroisemba, Imphal-795004, Manipur
E-mail: cau_dde@rediffmail.com

Introduction:

The State of Manipur lies between 23.8^{0} N to 25.68° N latitude and 93.03° E to 94.78° E longitude in the north –east corner of India. It has a pleasant and salubrious climate. The state has an area of 22,327 square kilometer; out of which 20,089 square kilometer is about 90 per cent of the total geographical area of the state are hilly and surrounded by the small central valley of 1,851 square kilometer .The total human population of Manipur in 2001 was 22.9 lakhs (0.22% of country's population) and the estimated population for the year 2011 is 27.93 lakhs as on 1st March, 2011. About 74.89 per cent of the population lives in rural areas. Schedule tribe and schedule caste population is 30.20 per cent and 2.77 per cent respectively. The altitude of the state above the mean sea level varies from 790 meters to 2020 meters. It has sub-tropical temperate climate.

Duck rising is one of the means for generating income especially to landless and marginal farmers. Rivulets, ponds marshy low lying lands filled with water due to high rainfall provide natural environment for raising ducks. The inhabitant on the brinks of Loktak Lake has a bright future for duck rearing. Khaki Campbell and non-descript type ducks are raised primarily for eggs Muscovy ducks are raised as a pet and for meat. White Pekin and Vigova Super M are also started introduction in Manipur.

Duck Population:

According to livestock census 2003, the total poultry population of Manipur was 29.41 lakhs consisting of 23.83 lakhs chicken and 5.33 lakhs ducks (Table-1). –1). Duckpopulation during 1972 and 1978 are 0.36 lakhs and 2.11 lakhs respectively, And, record of improved ducks is available from the year 1984 onwards only having 2.58 lakhs improved ducks which is 44.18% out of the 5.84 lakhs of total ducks. This may be attributed to the establishment of Duck Farm Thenguchingjin by the Department of Veterinary and Animal husbandry Veterinary and Animal Husbandry Services, Government of Manipur for popularization of Khaki Campbellducks in Manipur in the Year 1980.

Table1: Duck Population of Manipur (thousand)

Sl. No.	Item	1966	1972	1978	1984	1987	1992	1997	2003
1.	DESI								
	Drake								
	Duck	NA	13	86	87	89	93	71	81
	Duckling	NA	14	68	111	151	177	141	100
	(below 6 months of age)	NA	9	57	128	113	122	119	91
	Total Desi (% of total duck)	NA	36 (100%)	211 (100%)	326 (55.8%)	353 (54.6%)	392 (60.8%)	331 (55.9%)	272 (51.0%)
2.	IMPROVED								
	Drake	NA	NA	NA	72	67	51	56	71
	Duck	NA	NA	NA	81	131	113	107	90
	Duckling	NA	NA	NA	105	95	89	98	100
	(Below 6 months of age)								

	Total Improved (% of total duck)	NA	NA	NA	258 (44.18%)	293 (45.4%)	253 (39.2%)	261 (44.1%)	261 (49%)
3.	Total Duck (% if total Poultry)	NA	36 (3.6%)	211 (7.7%)	584 (20.4%)	646 (21.4%)	645 (19.8%)	529 (19.4%)	533 (18.1%)
4.	Total Poultry	790	1009	2748	2861	3021	3259	3057	2941

Desi duck population out of the total duck population as per livestock census report of 1984, 1987 and 1997 is about 55 per cent. It was recorded maximum during the year 1992 (60.8%) and minimum during the year 2003 (51%). The reverse is true in case of improved varieties. During 1984, 1987 and 1997 the population of improved duck out of the total duck population is about 44-45 per cent. It was minimum during 1992 (39.2%) and maximum during 2003 (49%). The result shows that there is increasing trend of the per cent improved ducks since 1992 to 2003.

Growth Pattern of Duck Population:

During 1972-1978 the annual growth rate of total duck population was 81 per cent and was maximum (Table – 2). Afterwards there is decreasing trend of annual growth rate. It follows negative trend from the year 1987 and the negative trend is maximum during 1997 to 2003 (-1.7%). The declining trend may be attributed to increase slaughtering of birds irrespective of their breed, sex and age. Ethnic violence in the state particularly in the hill districts before 1997 also contributed in the decrease of duck population. Non – availability of high yielding improved varieties of ducks in sufficient quantity and loss of local habitats may contribute to the negative trend of growth rate of the ducks.

Table 2: Growth Pattern of Duck Population (Annual Growth Rate %)

Sl. No.	Item	1966 to 1972	1972 to 1978	1978 to 1984	1984 to 1987	1987 to 1992	1992 to 1997	1997 to 2003
1.	Desi Duck	-	81	8.1	2.8	2.2	-3.1	-3.0
2.	Total Duck	-	81	29.5	3.5	-0.03	-1.6	-1.7
3.	Total Poultry	4.6	28.7		1.7	1.6	-1.2	-0.6

Egg Production:

Duck egg production in Manipur was 184 lakhs during 1992-93 (Table-3). The percentage of duck egg out of total egg production ranges from 25% to 27% during different years. There was negative trend of growth pattern of duck egg production from the year 1992-1993 upto 1995-96 (Table – 4) Positive trend of growth of egg production was observed from the year 1995-96 upto 2002-03 except 1999-2000 to 2000-01 in which it shows negative trend of -3.2%. The growth pattern was fluctuating from year to year. However, maximum duck egg production was recorded during the 2002-03 to 2003-04 (8.0%) and it may be due to the increased population of improved duck out of the total duck population (49.0%) as compared with other years. The growth pattern was lowest during 1993-94 to 1994-95 (-8-0%) and it coincides with the minimum number improved duck out of the total duck (39.2%) during 1992 census as compared with other years as presented in Table -1.

Layer farming is declining in Manipur as the cost of production is very high due to high transportation cost of feed from other states. The farmers cannot compete the cost of eggs coming from other states especially Andhra Pradesh. Every month about 14 truckloads of egg are importing to Manipur from outside states which amount to 454 lakhs eggs per year. This problem can be minimized by the popularization of rearing of low input technology birds and improved duck farming. Lakes and marshy lands of Manipur will favour popularization and expansion of duck farming.

Table 3: Estimates of Duck Egg Production.

Sl. No.	Item	1992-93	1993-94	1994-95	1995-96	1996-97	1997-98	1998-99	1999-2000
1.	Duck egg production								
	Desi	95	99	93	92	93	94	96	96
	Improved	89	76	68	67	68	68	78	89
	Total (% of total egg production)	184 (27.5%)	175 (25.9%)	161 (26.6%)	159 (26.8%)	161 (26.8%)	162 (26.3%)	174 (27.00%)	185 (27.1%)
2.	Total egg production (fowl & duck)	670	676	605	594	609	616	645	683
3.	Per capital availability of egg (Manipur) No./year/ person	-	27	28	30	30	31	32	33

Sl. No.	Item	2000-01	2001-02	2002-03	2003-04	2004-05	2005-06	2006-07
1.	Duck egg production Desi	94	97	95	105	113	107	107
	Improved	85	87	92	97	102	111	108
	Total (% of total egg production)	179 (25.8%)	184 (25.7%)	187 (25.4%)	202 (25.4%)	215 (26.2%)	218 (26.1%)	215 (25.7%)
2.	Total egg production (Fowl and duck)	695	717	736	794	822	835	836
3.	Per capita availability of egg (Manipur) No./year/ person	31	33	33	36	33	33	33

Table 4: Growth Pattern of Duck Egg Production (%)

Items	1992-93 to 1993-94	1993-94 to 1994-95	1994-95 to 1995-96	1995-96 to 1996-97	1996-97 to 1997-98	1997-98 to 1998-99	1998-99 to 1999-2000	1999-2000 to 2000-2001
Desi Duck egg	4.2	-6.1	-1.1	1.1	1.1	2.1	0	-2.1
Total Duck Egg	-4.9	-8.0	-1.2	-1.3	0.6	7.4	6.3	-3.2
Total Poultry Egg	0.9	-10.5	-1.8	2.5	1.1	4.7	5.9	1.8

Item	2000-01 to 2001-02	2001-02 to 2002-03	2002-03 to 2003-04	2003-04 to 2004-05	2004-05 to 2005-06	2005-06 to 2006-07
Desi duck egg	3.2	-2.1	10.5	7.6	-5.3	0
Total Duck Egg	2.8	1.6	8.0	6.4	1.4	-1.4
Total Poultry Egg	3.2	2.6	7.9	3.5	1.6	0.1

Meat Production:

Poultry meat production of Manipur for the year 2003-04 is 5253 M.T. (Table-5). This is the highest record during the different years. There is increasing trend of total meat production across different years and it was maximum during 2003-04 (22276 M.T.). Similarly duck meat production also is maximum during 2003-04 (951 M.T). There is declining trend of percentage of poultry meat out of total meat. This is due to the disproportionate increase in the proportion of meat from other sources except poultry.

Per capita availability of poultry meat show declining trend and is maximum during the year 1992-1993 (6.4 g/person/day) and minimum during the year 2003-04 (5.98g/person/day). The per capita availability of duck meat ranges from 1.1g/person/day to 1.3 g/person/day. There is declining trend from 1992-93 to 2003-04 and is maximum during 1992-1993 (1.3 g/person/day) and minimum during 2003- 04 (1.1g/person/day). It was 30.2%, 25.7% and 23.6% during 1992-93, 1997-98 and 2003-04 respectively.

Table 5: Estimates of Duck Meat Production (MT)

Item	1992-93	1997-98	2003-2004
Poultry Meat	4389	4642	5253
Total meat	14547	18047	22276
% of poultry meat out of total meat	30.2	25.7	23.6
Per capita availability of poultry meat g/person/day	6.4	6.0	5.98
Duck meat	869	901	951
Per capita availability of duck meat g/person/day	1.3	1.2	1.1

Per capita availability of poultry meat show declining trend and is maximum during the year 1992-1993 (6.4 g/person/day) and minimum during the year 2003-04 (5.98g/person/day). The per capita availability of duck meat ranges from 1.1g/person/day to 1.3 g/person/day. There is declining trend from 1992-93 to 2003-04 and is maximum during 1992-1993 (1.3 g/person/day) and minimum during 2003- 04 (1.1g/person/day). It was 30.2%, 25.7% and 23.6% during 1992-93, 1997-98 and 2003-04 respectively.

Constraints of Duck Development:

Even though the state is gifted with abundant flora and faun and majority of the people are non-vegetarian as well as traditional poultry lovers, commercial poultry farming is still at its infancy. The following constraints can be considered for duck development in Manipur.

(i) Non- availability of genetically superior ducklings:

Khaki Campbell ducklings are not available in sufficient quantity in Manipur and the farmers are facing difficulty for procurement of the same. As such the farmers are rearing local varieties with less production potential. Therefore, it is required to formulate feasible plan for production of commercial ducklings to meet the demands of the state as well as for neighboring states.

Identification of the various performance characters of the local varieties of ducks will be useful for preservation of indigenous germplasm as well as genetic improvement of the local varieties of ducks considering their adaptability to the agro climatic conditions of Manipur.

(ii) Non- availability of balanced feed and high:

(a) Non-availability of important feed ingredients like fish meat, ground nut cake, till oil cake, soya cake etc. in Manipur is a major bottleneck for computing quality poultry feed at compatible prices. Further transportation problem of the feed ingredients from outside the state of Manipur through hilly terrain during rainy season, economic blockade etc. make the feed cost dearer. Considering these problems, providing transport subsidy for transportation of poultry feed from outside the state of Manipur will be of great help to the farmers.

(b) At present there is no large sized feed manufacturer in Manipur and the farmers are compelled to depend on the feed procured from outside the state at costlier prices. Therefore, establishment of a large sized feed manufacturing mill at Manipur for manufacturing high quality poultry/duck feed will be of paramount importance in the interest of the poultry / duck farmers of Manipur.

(c) Research for locally available non-conventional poultry/duck feed ingredients for inclusion in poultry/duck feed will help to minimize the cost of the feed.

(d) During rainy season varieties of snails are available in plenty in marshy areas and lakes. They are very nutritious and useful for feeding the ducks. However, there is competition for human food. Research is required for identification of the various varieties of snails available in Manipur and large scale multiplication of the useful varieties. Processing and preservation of snail in dry form will be of immense help for manufacturing poultry/duck feed as a protein / mineral supplement.

(iii) Lack of marketing network:

A proper marketing network is required so that farmers/producers get the remunerative prices for their products and the consumer get hygienic poultry/duck products with consideration of the welfare of the birds.

(iv) Lack of adequate financial assistance from the financial institutions

Considering the poor financial condition of the farmers and reluctance on the part of the banks to finance the farmers on the ground of security problems for recovery of the loans, it is required on the part of the Government to formulate suitable schemes by which the farmers get loans without difficulty and also banks get their returns smoothly.

In this regard launching of Special Employment Generation program during the year 2003-04 by Manipur Government for self employment of youths aged between 20-45 years is a healthy development. The interest rate is low with 4 per cent per annum for loan amount up to Rs. 2 lakhs. The implementing agency of the program is Manipur State Co-operative Bank Ltd. Duckery is included in the program.

Proper planning and implementation for increasing the volume of duck rearing in Manipur will help to increase the economic condition of farmers and health status of the public and also to solve the unemployment problem Manipur is facing.

References:

Anonymous, 2000, Economic Survey of Manipur, 1999-2000. Directorate of Economics and Statistics, Government of Manipur, Imphal

Anonymous, 2002, Basic Animal Husbandry Statistics, Department of Animal Husbandry and Dairying, Ministry of Agriculture, Government of India, New DelhiAnonymous, 2003, 17th Quinquennial Livestock Census, Directorate of Veterinary & Animal Husbandry Services, Government of Manipur

Anonymous, 2004, Basic Animal Husbandry Statistics, Department of Animal Husbandry and Dairying, Ministry of Agriculture, Government of India,New Delhi

Anonymous, 2009, Economic Survey of Manipur, 2008-2009, Directorate of Economics and Statistics, Government of Manipur, Imphal

* * *

28

RURAL TOURISM-
A PRECURSER TO DEVELOPMENT
A CASE STUDY OF GARO HILLS

GINO A. SANGMA

ABSTRACT

Tourism is service-oriented sector which has made rapid strides globally in terms of gross revenue and foreign exchange earnings. The paper draws its strength from basic premise that tourism industry generates more employment opportunities (particularly in remote and backward areas) as well as helps in creation of necessary infrastructure like roads and telecom thus leading to overall development. This preamble highlighted by Antonio Fernandez-Morales, Maria Cruz Mayorga-Toledano and M. K. Sarma in their seminal work on tourism is further tested in the backdrop of Garo hills which is faster developing into a potential tourism hub.

Keywords: Tourism, Rural tourism and development

Introduction:

Tourism basically relates to travel for recreational, leisure or business purpose. Tourism has become a popular global leisure activity. The tourism sector stimulates other economic sectors through its backward and forward linkages and cross-sectional synergies with sectors like agriculture, horticulture, poultry, handicrafts, transport and construction. It contributes to the national integration; preserves natural and cultural environment; as well as enriches social and cultural lives of the people. It has the capacity to create substantial job opportunities, particularly for unskilled and semi skilled workers as well as to alleviate the poverty in the country.

Tourism gives a financial incentive to preserve architectural heritages and helps the survival of art forms, crafts and culture. It is a composite of service providers, both public and private, which includes travel agents and tours operators; air, rail, and sea transportation operators; guides; owners of hotels and guest houses and inns, restaurants and shops; etc.

Conceptual Framework and Review of Literature:

Tourism, in general, is a sector that can stimulate development. Although we do not consider that it is a universal solution to all the difficulties of the developing countries and whereby it has also many positive effects on development, where it creates commercial activities and a new market, increases urbanization, increases the state earnings and bridging or reducing whatever deficit there is in balance of payments and thus activates the economic circuit in a country (Wahab 1997).

Rural Tourism, in particular, can contribute to economic growth, heritage conservation and enhancement, and achievement of a higher standard of living for the inhabitants of rural areas. It can foster the creation of opportunities for local development through job creation and local projects for the conservation of animal species, plants or biodiversity in general. Heritage (in a broad sense, including natural and cultural elements) can be a strategic resource for the socio-economic development for local people, as in many cases it represents the basic component of rural tourism that drives and sustains small-and medium-sized enterprises. This is the reason why tourism's multiplier effect has a positive influence on the different sectors of regional economies. Rural tourism can also stimulate greater environmental awareness and instill a stronger sense of pride among residents, when they see tourists enjoying their own historical, cultural and natural heritage (Fernandez and Mayorga 2008).

Definition of the World Tourist Organization (WTO): **Tourists** are people who are "travelling to and staying in places outside their usual environment for not more than one consecutive year for leisure, business and other purposes not related to the exercise of an activity remunerated from within the place visited" (Uherek 2008).

Tourism Development basically relates meeting a growing demand from tourism poses some critical challenges. It is argued that there are three main areas which policymakers need to be concerned with: infrastructures, education and safety. The development of infrastructure is vital in any tourism project such as: household utilities as water, electricity and telephone and transportation facilities as roads or public transport system and other kind of tourist infrastructures that may help to promote tourism to the region (Gee et.al. 1989). This one will be used as operational definition for the proposed study.

The marketing experts opine that tourism marketing is the systemic and coordinated efforts to optimize the satisfaction of tourism. The tourism marketing is also supposed to be a device to make a possible reorientation in the business policy and overhaul in the management concept. Generally speaking, tourism planning has been defined as a process based on research and evaluation, which seeks to optimize the potential contribution of tourism to human welfare and environmental quality (Tosun and Jenkins 1998).

Kotler (2003) developed a new concept of Developmental Marketing which explains about how to convert latent demand into actual demand through designing and implementing appropriate marketing mix strategies for tourism development.

A tourist destination may be defined as a country, state, region, city or town which is marketed or markets itself as a place for tourists to visit (Chawla 2008). It is basically a city, town, or other area that is dependent to a significant extent on the revenues accruing from tourism. It may contain one or more tourist attractions and possibly some "tourist traps" (Gunn 1997).

A tourist attraction is a place of interest where tourists visit, typically for its inherent or exhibited cultural value, historical significance, natural or built beauty, or amusement opportunities. Tourist attractions are also created to capitalize on legends such as a supposed UFO (Unidentified Flying Object) crash site near Roswell, New Mexico. Ethnic communities may become tourist attractions, such as Chinatowns in the United States and the Black British neighborhood of Brixton in London, England (Conde Nast Traveler Magazine, Nov 2000).

In recent times Destination Management is gaining increasing attention from National Tourism Organisations (NTOs). It is also acknowledged that image management is a very important area of overall destination management. The concept of image management consists of building and maintenance of external image of a destination. The image of a destination may be tourist segment specific (Kotler et. al. 2004).

Destination image, defined as the compilation of beliefs and impressions based on information processing from various sources over time (Crompton 1979)

Destination positioning, which is a process of building and maintaining images for a destination helps in developing and implementing particular strategies of the marketer (Payne 1996).

Ries and Trout argue that for creating distinctive positioning in the tourist mind, tourism marketers have to follow either of the three strategic alternatives such as: strengthening its own current position or; to grab an unoccupied position or; to deposition or reposition their products. According to Michael Treacy and Fred Wiersema, marketers can position their tourist products in either of any of the three ways: a firm could aspire to be the product leader or; the operationally excellent firm or; the customer intimate firm (Kotler 2003).

An optimal tourism development might take into account the current level of knowledge of the population and the speed of its potential improvement. This is relevant in order to design an optimal tourism development plan over time. Safety is usually a highly appreciated feature in tourism resorts. Most of the tourists look for places to spend a nice and non-problematic stay. In this sense, it is argued that most of the tourists are risk averse. This is a very important issue to take into account when assessing the competitiveness of a tourist destination (Gee, et.al. 1989).

Tourism planning should relate tourism development to the more equitable distribution of wealth that is one of the main aims of national development planning. In this respect, tourism planning is a component of national development planning and strategy. Moreover, it includes a decision-making process between the tourism industry and other sectors of the economy, between various sub-national areas and between types of tourism. It requires the integration of the tourism industry into other sectors such as agriculture, industry, transportation and social services (Timothy 1998).

Tourist destinations being one of the major products associated with the tourism industry are also no exceptions. Destinations may develop image(s) without conscious efforts of the marketers due to the fact that these forms of products are widely discussed, debated and visited by the customers (Sarma 2000). Many developing countries have prepared plans particularly at the central level to guide tourism development, as they have recognized the tourism sector as an important source of foreign currency earning and employment (Tosun and Timothy 2001).

It is well established that every product generates images among the target customers and this image is known as position of the particular product. The position enjoyed by the same product may be different for different customer groups. Positioning might also give the desired impetus to a destination, as the prevailing word of mouth may not supply the market with "the material to talk about" (Ries & Trout 1986). The concept of positioning of tourist destinations may be successfully used for some other tasks as well. Repositioning may help in lengthening the maturity phase or initiating rejuvenation of the destination life cycle (Gilbert 1990).

Destination positioning, primarily which focuses on the process of building and maintaining images for a destination helps in developing and implementing particular strategies of the marketer. Extant of positions may not be desirable from the point of view of the concerned marketers; or such position(s) may not be sufficient to achieve desired marketing goals. It necessitates a detailed scientific study on the level of preference and perception of particular group(s) of tourists. These preferences can then be promoted for the destination in question and if needed a new and profitable position may be developed as a result (Payne 1996). Elements like destination choice, image, and satisfaction have all been the subject of considerable tourism research (Sirakaya et.al. 1996).

Murphy (1999) has identified the "image" of a destination as very important as far as promotion is concerned; it also aimed at tourism marketing through positioning. He also concluded that such studies tried to match the needs of one or more market segment(s) with those actually present in the destination by promotional campaign. Though the concept of positioning is by now more than two decades old and quite popular among manufactured goods and other services marketers, it is yet to be used as a successful marketing strategy in promoting considerable number of tourist destinations (Sarma 2007).

Destination image and its measurement have long been on the research agenda because by creating a favorable image a destination will attract tourists and achieve profitability (Phelps, 1986). Destination marketers have been trying to catch the consumers' leisure time and disposable income relentlessly. Interestingly, unlike other products and services, there is nothing urgent or emergency (under normal circumstances) about a holiday. Therefore, the panacea of leisure marketing lies, among others, in promoting and communicating the image of the product, mainly that of the destination. Importance of image in creating acceptability among the target market and concluded that destination image plays crucial role in travelers' purchase related decision making (Sarma 2003).

Pike (2002) have reported that studies on destination image suggest successful achievement of the goal of matching the needs of the target groups with the physical features actually present in destination. Like all products the tourist destinations also vie for the limited time and resources available at the disposal of the consumers (tourists). Tourists too build a psychological choice set of competing destinations before finally selecting a place to visit. Then the complicated process of elimination is initiated, where the tourist mentally matches the desire with perceived offerings of individual destinations. During this matching process the average tourist takes up certain factors with her own preference levels. Sarma (2003) noted that whichever destination matches the preference of the tourist to the closest extent would be finally taken up for the visitation.

Factors affecting destination choices:

The researchers over the years have identified various factors like safety, tourist attraction, local people/culture, weather, cost of accommodation and transportation, transportation to the destination, etc are responsible for destination choices. The three broad phases of tourism namely, the dynamic phase, a static one, and the consequential give rise to different ingredients of the industry. These ingredients can be grouped into three macro level components namely, (a) Transportation (Dynamic phase), (b) The Destination (Static phase), and (c) Range of facilities in the destination like accommodation and services of guide etc. (a combination of Dynamic and Static phases). All the three factors individually and collectively affect the satisfaction level of the visitors have identified different micro level factors those influence the destination choice of the tourist several other researchers (Robinson 1976; Archer 1976).

The elements like safety, pleasure, facilities available etc. could individually influence the tourist's desirability of the destination. It must be pointed out here that tourism being a service, the amount of information or knowledge the traveller is going to gather on a destination is very crucial; which might even influence the whole decision making process (Zeithaml 1981). Mathieson and Wall (1982) also identified these three main phases of the tourism industry and have suggested that various factors individually and collectively affect the satisfaction level of tourists.

Godfrey & Clarke (2000) emphasized that it is possible to exploit each of these factors for marketing a destination. These three phases mentioned above may be regarded as the broad product mix of a tourism destination as it is possible to modify and improvise these elements in order to implement a particular marketing strategy.

Influence of information sources on tourists destination choice:

Phelps (1986) reported that the prospective tourists who may have limited or incomplete information (on comparable destinations that would constitute the primary choice set) at his/her disposal are able to decide about a destination without having full or reasonable information. He used structured list consisted of scenic beauty (of Menorca--the Spanish island, British resorts etc.), climate, and cultural/historical attributes. Gartner (1993) again described the roles of different components like traditional advertising, information from professional tour operators, and earlier visitors' recommendation in his study of choice of destinations.

Murphy (1999) used few other variables like safety, travel cost, and local people to measure tourists' perception regarding Australia. Obviously, this can be done through medium of communication those serve as the sources of information for the tourists. Understanding of the variables is always of an utmost importance in consideration of destination choice.

Accessibility has always been a factor for destination choices, as an average tourist will not like to visit a place, which is not easily accessible. Likewise, accommodation, tourist attractions and climate of the destination during the time of visit are also equally important to a tourist. A prospective tourist will also weigh the infrastructure available in the would-be-destination in many forms including that of availability of drinking water. Word-of-mouth and directed campaign from the respective NTO (National Tourism Organization) through the distribution conduit might also have a meaningful role to play (Sarma 2007).

Destination management:

Destination management facilitates the destination's positioning and thereby reduces the risk elements for the prospective travelers, especially for those who are visiting the destination for the first time (Kamra 1997). Creation of new images and subsequent switching among them also provide the marketers with the discretion of choosing target market(s) (Baum, 1998). Image management helps a destination management in lengthening the maturity phase and initiation of rejuvenation (both crucial stages of the destination life cycle) of a destination (Andriotis 2001). Pike (2002) reported 142 literatures related to destination image, which indicate the present day researchers' interest on destination image building and management.

It has widely been reasoned that the hitherto underdeveloped and politically disturbed local economy can be flourished only through grass-root level developments like those are possible through tourism (Sarma 2003). In recent times destination management has gain its interest and is also acknowledged that image management is considered to be a very important area for overall destination management (Kotler et.al. 2004).

In view of the aforesaid, facts it is right to mention that tourism is an integrated effort to satisfy tourists by making available to them the best possible services. Marketing process will thus act as device to transform the potential tourists into actual tourists. The above Review of Literature revealed that most of the studies were conducted in various developed and other developing countries. A very few studies are conducted in North-East India. Especially no study of this nature has been conducted in Garo Hills where lot of tourism potential is not harnessed leading to a research gap needed to be bridged.

Objectives of the Study:

Based on literature review and the research issue identified above, researcher has converged to following objectives:

1. To study the important features of tourist attractions in Garo Hills.
2. To develop tourism marketing strategies and approaches towards positioning Garo Hills as tourist destination.

Major Findings

Garo hills as a case was taken up to highlight a comparative development picture over a period of five years (2006-2010) wherein a sustained development of tourism is taken up. The data sources analysed were basically secondary and few interviews were also conducted of the government and other officials involved in tourism promotion and its sustenance.

Garo Hills: A Brief Profile

North East India is situated in the eastern most corner of India surrounded by the Himalayas, the plains of Bangladesh and the hill ranges of Myanmar. The North-East region is full of scenic beauty, which can attract tourists of almost any taste and represents different unique ethnic heritage and culture. The region houses a large number of tribes rich in their unique culture, heritage and rituals, and with a huge number of dialects unique to the tribes (RCILTS, 2002). The lush green semi-tropical forests of the region house, besides others, the famous one-horned Asiatic rhino, wild buffalo, and elephants. Many other endangered species of flora (including a wide variety of orchids) and fauna are also found in the forests of the region.

The Garo Hills are the part of range that is called the Garo-Khasi range in Meghalaya. They are inhabited mainly by tribal dwellers. Shillong, the capital of Meghalaya, is located in this range. It is one of the wettest places in the world. The range is part of the Meghalaya subtropical forests eco-region. The Garo Hills have an area of approximately 8,000 sq. Kms. Garo Hills comprises 3 (three) districts, Garos predominantly inhabit the three district of the State namely East Garo Hills, West Garo Hills and South Garo Hills. Tura is the largest town with a population of about 70,000 located at the foothills of often cloud covered Tura peak. The localities have their own customs and traditions very different from city dwellers and also indulge in superstitious acts.

Through the Garo Hills, two mountain ranges pass, called the Arbella range and the Tura range, which forms the great Balpakram valley in between. The town is centrally located to other popular game/wild life sanctuaries in the district such as Balpakram and Nokrek, natural caves (the Siju cave being one of the longest in Asia). These places are rich reserves of natural flora and fauna. From the Tura peak one can get the amazing view of the sunset at 1,400m and its peak can reach by a 5 km trek, by hiking or rock climbing.

The Nokrek Peak is the highest point in the Garo Hills, which is covered by thick film of lust forest. Flora and Fauna is in plenty in the hills. The mother of germo plasm 'Citrus indica' rare specie has been discovered by science researchers within Nokrek Range. This discovery led to the establishment of the National Citrus Gene Sanctuary-cum-Biosphere Reserve here. The Nokrek National Park comprises herds of wild elephants, many rare varieties of birds and pheasants along with rare orchids.

Balpakram area is considered to be sacred for the Garo community due to their belief that the spirits of the dead sojourn here. Balpakram is known as the land of perpetual winds, it is one of the most popular tourist spot of Garo Hills. The plateau has enchanting views of the plains of Bangladesh with meandering rivers and large expanses of inland water. Herbs and plants of medicinal value can be found here. There is a great precipice or deep gorge in Balpakram and is popularly compared to the Grand Canyon of U.S.A.

Garo Hills has variety of attractions to offer to the tourists. Other major tourist spots are, Siju cave, Nengkong cave, Siju Bird Sanctuary, Nengkong, Tura peak, Rongbang Dare, Pelga Fall, Bhaitabari, Wadagokgre Imilchang Dare, Adokgre, Teteng Cave, Sasatgre Village, Williamnagar, Sisobibra, Baghmara Reserve Forest, etc. are some of the places that has always been centre of attraction in the Garo Hills region for tourism.

Garo Hills, festivals include Wangala Festival the harvest festival, which takes place every November at the hills. Hence the Garo Hills with its vibrancy and vigour can culminate into an interesting holidaying experience. One can feel at paradise with the greenery and the fresh air, which might help to rejuvenate and recuperate. The tourism infrastructure under progress is mentioned below:

List of schemes sanctioned for Garo Hills by the Ministry of Tourism, Govt. of India.

Name of the Project	Nature of Work	Amount Sanctioned
Destination development at Tura 2006 – 2007	Tura Orchid Lodge additional resort, wayside amenities, picnic spot, viewpoints etc.	Rs. 487.00 Lakhs
Development of circuit Tourism Barengapara-Gasuapara-Rongara-Mahadeo-Balpakram 2008-09	Wayside amenities, tourist lodge, viewpoint, toilet, footpath, site development, boat house, etc	Rs. 642.85 Lakhs
Development of Rural Tourism in Sasatgre Village (Hardware)	Traditional House, Nokpante, crafts shop, toilet, car parking, etc.	Rs. 38.49 Lakhs
Software Work Plan under capacity building for service providers schemes at Sasatgre, West Garo Hills.		Rs. 23.95 Lakhs
Development of circuit Tourism Tura-Garobadha-Selsella-Bhaitabari-Phulbari-Tikrikilla 2009-10	Wayside amenities, tourist lodge, viewpoint, toilet, footpath, site development, preservation of archaeological site, boat facilities, etc	Rs. 538.23 Lakhs
Destination development at Resubelpara, East Garo Hills 2009 – 2010	Tourist Lodge, toilet, cooking shed, boat house, rain shelter, sit-out, footpath, preservation of fossils, phyller stones etc.	Rs. 281.11 Lakhs

(***Source:*** INTACH, Tourism Development Plan for Meghalaya, 2010

Conclusions and Suggestions

Rural Tourism appears to be the better option than mass tourism for the development of economically depressed areas, as the required dimension of businesses allows rural population to enter in the sector. This facilitates the participation of local communities in sharing benefits and decision making (Tosun 2004).

Most importantly domestic market dominates rural tourism in Garo Hills, as it occurs in many developing countries. Therefore, in order to attract international and domestic tourists for this segment the positioning of the product in the international and domestic market would be strengthened by the coordination of promotion amongst the different regions or even countries. The public may help in the development of this sector through promotion, marketing and research into market and trends.

Considering the special characteristics of the natural and social environment in which rural tourism takes place, it is necessary to protect it and to guarantee the future sustainability of the sector through the establishment of an adequate and effective legal framework, before a disordered growth with irreversible consequences occurs. It is also very important that the legal framework does not impose complex administrative processes that slow down the development of rural tourism, and that a certain degree of coordination with neighbouring regions or counties exists.

It is also suggested that active policies, like promotion, training, institutional reinforcement and marketing are essential for a development of rural tourism. The increasingly stressful urban settings provide an opportunity to urban dwellers with a recreational space amidst rural landscapes with natural and cultural diversity. In a way, tourism in rural areas can be categorized as cultural tourism where a visitor seeks to enhance personal experience through exchange of knowledge, interactions with different cultures and environment. Rural tourism can also form the base for eco-tourism, heritage tourism or simply recreation tourism.

References:

Andriotis, K. (2001). "Strategies on Resort Area and Their Life Cycle Stages", *Tourist Review*, 56 (1/2), 40-43.

Archer, B. H. (1976). *Occasional Papers in Economics*. Bangor: University of Wales Press

Baum, Torn (1998). Taking the exit route: extending the tourism area life cycle model. Current Issues in Tourism, 1(2), 167-175.

Chawla, R (2008) "Tourism Marketing and Crisis Management", Rajat Publications, New Delhi.

Conde Nast Traveler Magazine (November, 2000). *Reader's choice awards 2000*. New York: Conde Naste publications Ltd.

Crompton, J. L. (1979). Motivations for pleasure vacation. Annals of Tourism Research, 6(4), 408-424. Echtner, C. M., & Ritchie, J. R. B. (1991). The meaning and measurement of destination image. The Journal of Tourism Studies, 2(2), 2-12.

Fernandez-Morales, A and Mayorga-Toledana, M.C. 2008. 'Rural Tourism in Andalucia: Recent evolution and normative regulation', in Sutheeshna Babu. S, Mishra, S and Parida, B.B (eds), Tourism Development Revisited: concepts, issues and paradigm, pp.105-119. Los Angeles: Response.

Gartner, W. (1993). Image formation process. In Uysal, M., & Fesenmair, D. (Eds.), Communication and Channel System in Tourism Marketing. New York: Haworth.

Gee, C. Y, Makens, J.C. and Choy, D. (1989), *The Travel Industry*, 2nd ed. New York: Van Nostrand Reinhold.

Godfrey, K and Clarke, J. (2000), *The Tourism Development Handbook*. New York: Cassell.

Gilbert, D. (1990). Strategic marketing planning for national tourism. Tourist Review,45(1), 18-27.

Gunn, C. A. (1997), *Vacationscape: Developing Tourist Areas*, 3rd ed. Washington, D. C: Taylor & Francis.

INTACH, Tourism Development Plan for Meghalaya (April 2010).

Kamra, K.K. (1997). Tourism: Theory, Planning and Practice (pp. 82). New Delhi: Indus Publishing Company.

Kotler, Philip (2003). Marketing Management, *eleventh edition*, Pearson Education (Singapore) Pte. Ltd., Indian Branch, 482 F.I.E. Patparganj, Delhi, India

Kotler, P., Bowen, J. & Makens, J. (2004). "*Marketing for Hospitality and Tourism*" (pp. 744-745). New Delhi:Pearson.

Matheison, A., & Wall G. (1982). Tourism: Economic, Physical and Social Impacts. Harlow: Longman.

Murphy, Laurie (1999). Australia's image as a holiday destination - perception of backpacker visitors. Journal of Travel and Tourism Marketing, 8 (3), 21-45.

Payne, A. (1996). The Essence of Services Marketing (pp.109). New Delhi: Prentice Hall of India Private Limited.

Phelps, A. (1986, September). Holiday destination image-the problem of assessment. An example developed for Menorca. Tourism Management, 168-181.

Pike, S. (2002). "Destination Image Analysis -- A Review of 142 Papers From 1973 To 2000", *Tourism Management*, 23, 541-549.

Ries, Al, & Trout, Jack (1986). Positioning: The Battle of Your Mind. New York: McGraw-Hill International.

Robinson, H. (1976). *A Geography of Tourism.* New York: McDonald and Evans.

Sarma, M. K. (2000). Destination positioning of north east India based on preferences and perceptions of tourists. Unpublished doctoral dissertation, Tezpur University, Tezpur.

Sarma, M.K. (2003). "Positioning A Tourist Destination: A Study of North East India", *ASEAN Journal on Hospitality and Tourism*, 2(2), 104-117.

Sarma, M.K (2007) Influence Of Information Sources on Tourists: A Segment-wise Analysis with Special Focus on Destination Image *Vision: The Journal of Business Perspective, Volume 11 Number 1*

Sirakaya, E., McLellan, R.W. and Uysal, M. (1996), "Modeling vacation destination decisions: a behavioral approach", Journal of Travel and Tourism Marketing, Vol. 5 No. 1-2, pp. 57-75.

Timothy, D. (1998), "Cooperative tourism planning in a developing destination", Journal of Sustainable Tourism, Vol. 6 No. 1, pp. 52-68.

Tosun, C. and Jenkins, C.L. (1998), "The evolution of tourism planning in third world countries: a critique", Progress in Tourism and Hospitality Research, Vol. 4 No. 2, pp. 101 14.

Tosun, C and Timothy, D.J "Shortcomings in planning approaches to tourism development in developing countries: the case of Turkey" International Journal of Contemporary Hospitality Management 13/7 [2001] 352.

Uherek, E (2008), Tourism in Europe and in the world, retrieved from http://www.atmosphere.mpg.de/enid/Nr_9_July__6_Air_traffic/C__Tourism_5rw.htm, on 15th July, 2010.

Wahab, S. 1997. 'Sustainable Tourism in the Developing World', in S. Wahab and J. P. Pigrim (eds), *Tourism, Development and Growth*, pp. 126-46. London: Routledge.

Zeithaml, Valarie A. (1981). Marketing of Services. Chicago: American Marketing Association.

* * *